Name: _____

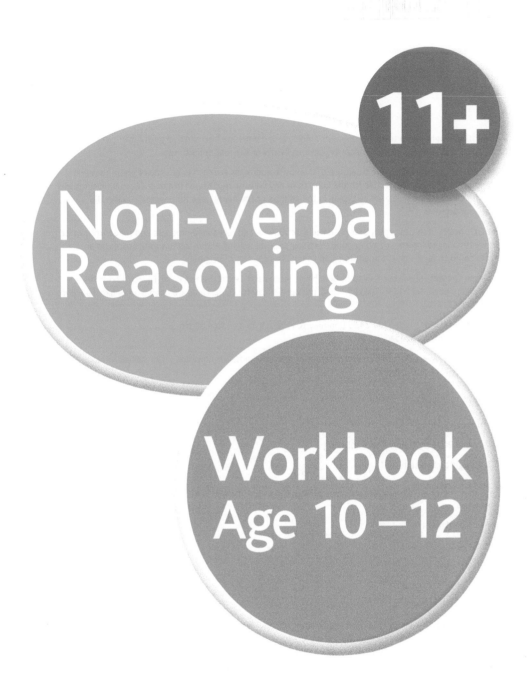

11+

Non-Verbal Reasoning

Workbook
Age 10–12

Alison Primrose

GALORE PARK

AN HACHETTE UK COMPANY

Orders: please contact Hachette UK Distribution, Hely Hutchinson Centre, Milton Road, Didcot, Oxfordshire, OX11 7HH. Telephone: (44) 01235 400555. Email: primary@hachette.co.uk. Lines are open from 9 a.m. to 5 p.m., Monday to Friday.

Parents, Tutors please call: 020 3122 6405 (Monday to Friday, 9:30 a.m. to 4.30 p.m.). Email: parentenquiries@galorepark.co.uk

Visit our website at www.galorepark.co.uk for details of revision guides for Common Entrance, examination papers and Galore Park publications.

ISBN: 978 1 4718 4936 7

© Alison Primrose 2016

First published in 2016 by Galore Park Publishing

Hodder & Stoughton Limited

An Hachette UK Company

Carmelite House

50 Victoria Embankment

London EC4Y 0DZ

www.galorepark.co.uk

Impression number   10 9 8

Year   2024 2023 2022

Illustrations by Peter Francis.

The following additional illustrations by Integra Software Services Ltd:

p.17 bottom left, p.21 bottom, p.A1 right, p.A2, p.A3, p.A4 top left, p.A8 bottom left and middle right

Typeset in India

Printed in the UK

A catalogue record for this title is available from the British Library.

Contents and progress record

Use these pages to record your progress. Colour in the boxes when you feel confident with each skill and note your scores for the 'Test yourself' and workout questions.

How to use this workbook

Introduction

This workbook has been written to help you develop your skills in Non-Verbal Reasoning. The questions will help you:

- learn how to answer different types of questions
- build your confidence in answering these types of questions
- develop new techniques to solve the problems easily
- practise maths skills that can help improve your abilities in Non-Verbal Reasoning
- build your speed in answering Non-Verbal Reasoning questions towards the time allowed for the 11+ tests.

Pre-test and the 11+ entrance exams

The Galore Park 11+ series is designed for pre-tests and 11+ entrance exams for admission into independent schools. These exams are often the same as those set by local grammar schools too. 11+ Non-Verbal Reasoning tests now appear in different formats and lengths and it is likely that if you are applying for more than one school, you will encounter more than one of type of test. These include:

- pre-tests delivered on-screen
- 11+ entrance exams in different formats from GL and CEM
- 11+ entrance exams created specifically for particular independent schools.

Tests are designed to vary from year to year. This means it is very difficult to predict the questions and structure that will come up, making the tests harder to revise for.

To give you the best chance of success in these assessments, Galore Park has worked with 11+ tutors, independent school teachers, test writers and specialist authors to create this series of workbooks. These workbooks cover the main question types that typically occur in this wide range of tests.

For parents

This workbook has been written to help both you and your child prepare for both pre-test and 11+ entrance exams.

The content doesn't assume that you will have any prior knowledge of Non-Verbal Reasoning tests. It is designed to help you support your child with simple exercises that build knowledge and confidence.

The exercises on the **learning spreads** can be worked through either with your support or independently. They have been constructed to help familiarise your child with how a question type works in order to build confidence in tackling real questions.

The **maths workout** sections are provided to help consolidate learning in related areas of maths.

This workbook has been written for teachers and tutors working with children preparing for both pre-test and 11+ entrance exams. The wide variety of question types is intended to prepare children for the increasingly unpredictable tests encountered, with a range of difficulty developed to prepare them for the most challenging paper and on-screen adaptable tests.

Working through the workbook

- The **contents and progress record** helps you keep track of your progress. Complete it when you have finished one of the **learning spreads** or **maths workout** sections.
  - Colour in the 'Completed' box when you are confident you have mastered the skill.
  - Add your 'Test yourself' scores to track how you are getting on and to work out which areas you may need more practice in.
- **Chapters** link together types of questions that test groups of skills.
- **Learning spreads**, like the one shown here, each cover one question style.

Have a go

Try these activities to build your skills towards answering the exam-style questions.

Test yourself

Complete a set of exam-style questions that includes some challenging problems.

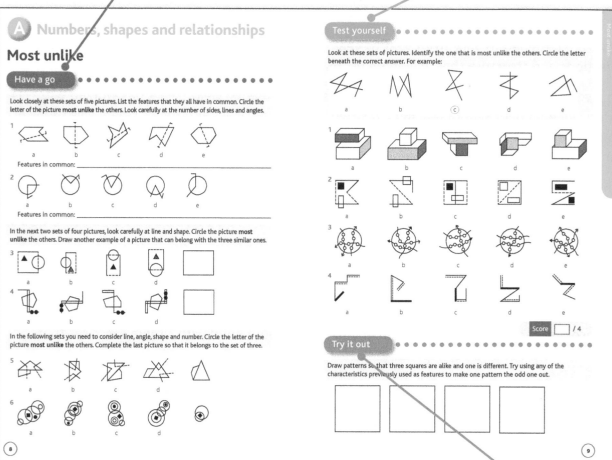

Try it out

Use your new skills to create your own questions or complete a fun activity.

- **Maths workouts** help you to practise familiar skills that link to the Non-Verbal Reasoning questions in this workbook.
- **Answers** to the **Have a go**, **Test yourself** and **Try it out** questions can be found in the middle of the workbook. Try not to look at the answers until you have attempted the questions yourself. Each answer has a full explanation so you can understand why you might have answered incorrectly.

Test day tips

Take time to prepare yourself the day before you go for the test: remember to take sharpened pencils, an eraser and a watch to time yourself (if you are allowed – there is usually a clock present in the exam room in most schools). Take a bottle of water in with you, if this is allowed, as this will help to keep your brain hydrated and improve your concentration levels.

... and don't forget to have breakfast before you go!

Continue your learning journey

When you've completed this workbook, you can carry on your learning right up until exam day with the following resources.

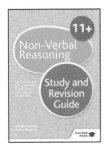 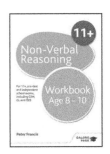 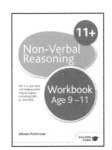  

The *Study and Revision Guide* introduces the skills and question types you may encounter in your Non-Verbal Reasoning 11+ entrance exams. Questions are broken down into the familiar maths areas you have learned about in school, with full explanations to show how these work together in the Non-Verbal Reasoning questions.

The workbooks will develop your skills, with many practice questions. To prepare you for the exam, these books include even more question variations that you might encounter. The more question types you practise, the better equipped for the exams you'll be. All the answers are explained fully.

Workbook Age 8–10: Increase your familiarity with variations in the question types.

Workbook Age 9–11: Experiment with further techniques to improve your accuracy.

The *Practice Papers* (Books 1 and 2) contain four training tests and nine model exam papers, replicating various pre-test and 11+ exams. They also include realistic test timings and fully explained answers to help your final test preparation. These papers are designed to improve your accuracy, speed and ability to deal with a wide range of questions under pressure.

Numbers, shapes and relationships

Most unlike

Have a go •

Look closely at these sets of five pictures. List the features that they all have in common. Circle the letter of the picture **most unlike** the others. Look carefully at the number of sides, lines and angles.

1

  a   b   c   d   e

Features in common: _____

2

  a   b   c   d   e

Features in common: _____

In the next two sets of four pictures, look carefully at line and shape. Circle the picture **most unlike** the others. Draw another example of a picture that can belong with the three similar ones.

3

  a   b   c   d

4

  a   b   c   d

In the following sets you need to consider line, angle, shape and number. Circle the letter of the picture **most unlike** the others. Complete the last picture so that it belongs to the set of three.

5

  a   b   c   d

6

  a   b   c   d

Test yourself

Look at these sets of pictures. Identify the one that is most unlike the others. Circle the letter beneath the correct answer. For example:

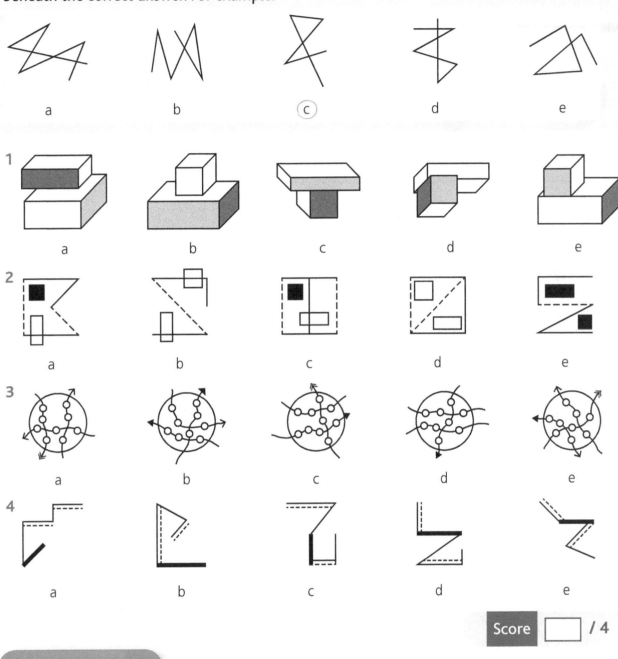

Score [ ] / 4

Try it out

Draw patterns so that three squares are alike and one is different. Try using any of the characteristics previously used as features to make one pattern the odd one out.

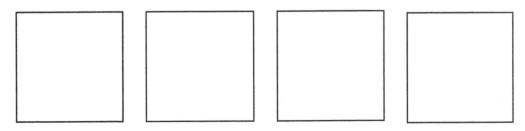

Matching features 1

What do the following pairs of pictures have in common?

1

Features in common: _____

_____

_____

2

Features in common: _____

_____

_____

In what ways are the following pairs of pictures different from each other?

3

Differences: _____

_____

_____

4

Differences: _____

_____

_____

In the next two questions identify the common features between the pairs and draw another picture to match the pair in the space provided.

5   6

Test yourself

Look at the first two pictures and decide what they have in common. Then select one of the options from the five on the right that belongs in the same set. Circle the letter beneath the correct answer. For example:

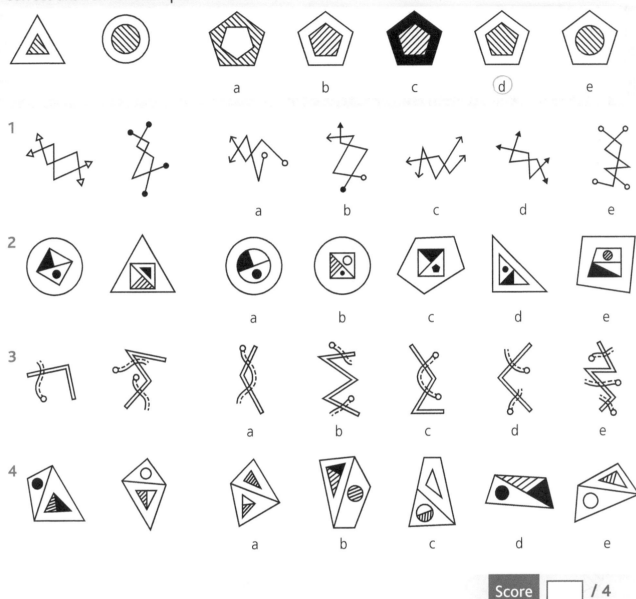

Try it out

Choose one of the patterns already drawn on the right and draw two patterns on the left that will belong with the one you have chosen. None of the others should belong to the group of three you have created.

a   b   c   d   e

Applying changes 1

Work out how the first pattern has changed to give the second pattern. Then draw a fourth pattern, which is formed when the third pattern is changed in the same way as the first pair.

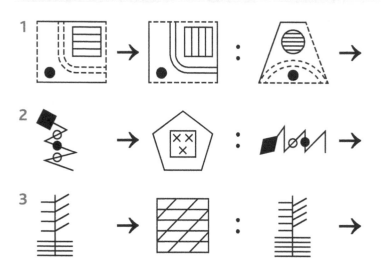

Add shading to the empty fourth pattern so that the third pattern is changed in the same way as the first pattern was changed to form the second pattern.

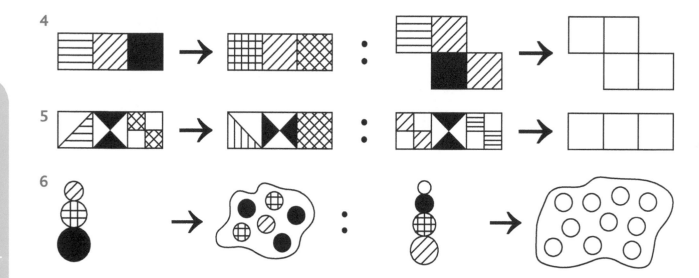

Test yourself

Look at the two pictures on the left connected by an arrow. Decide how the first picture has been changed to create the second. Now apply the same rule to the third picture and circle the letter beneath the correct answer. For example:

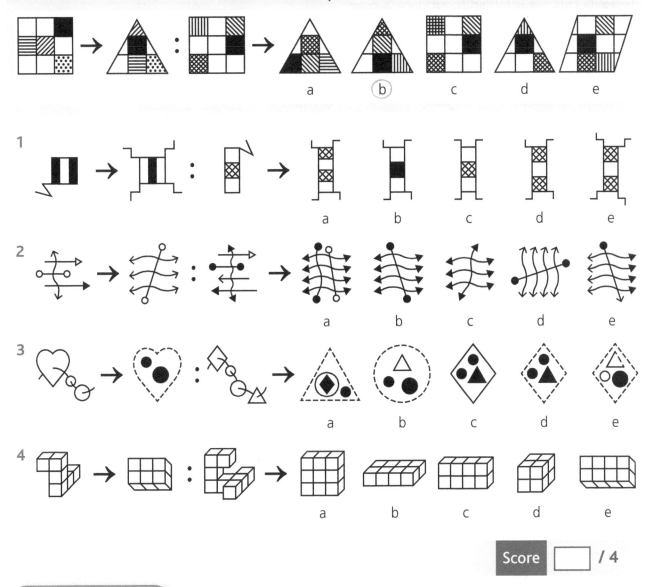

Try it out

Think of a way to change each of these patterns and draw the resulting second pattern. Then draw a third pattern and change it in the same way as the first pair to give the fourth pattern.

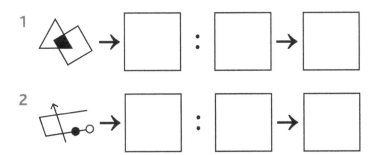

Matching 2D and 3D shapes 1

Imagine this net is folded to produce a cube:

1

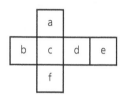

Which face will be opposite:

(a) face c? _____

(b) face a? _____

(c) face d? _____

2 When folded, which net will produce this 3D shape? Circle the letter beneath the correct net.

  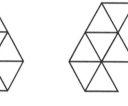  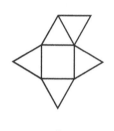

        a   b   c   d   e

The next two nets have been folded to produce a cube. Draw the missing pattern on the front face.

3

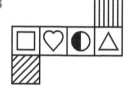

4

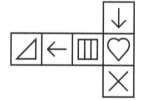

In the next two questions draw the missing pattern on the net produced for these cubes.

5

 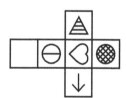

6

 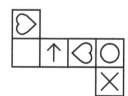

Test yourself

Find the cube, or other 3D shape, that can be made from the net shown on the left. Circle the letter beneath the correct answer. For example:

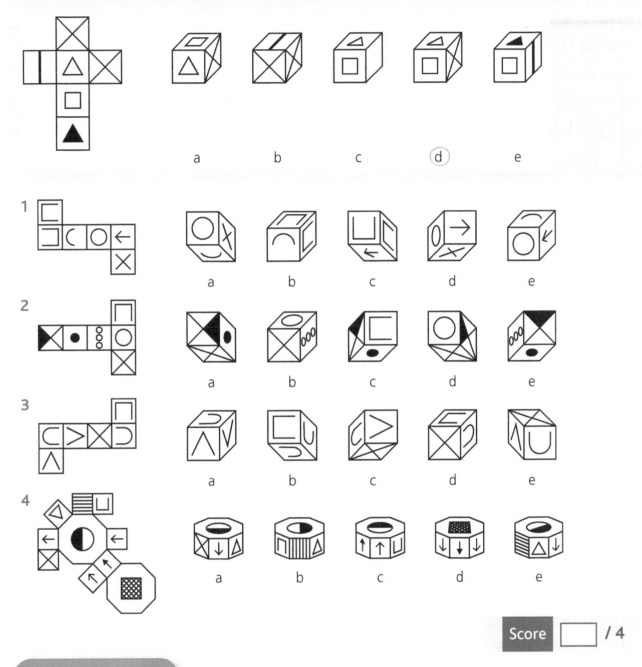

a   b   c   (d)   e

Score [   ] / 4

Try it out

Draw patterns or shapes on the net. Then draw on the faces of the folded cubes so that two of the three cubes are correct and one cube **cannot** be made from the folded net even though it has the same patterns on it.

1

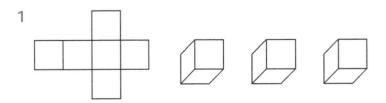

Matching features 2

List the features these sets of pictures have in common and then draw another member of the group containing these features.

1

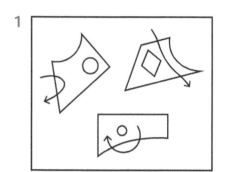

Features in common:

_____

_____

_____

_____

Another member of the group:

2

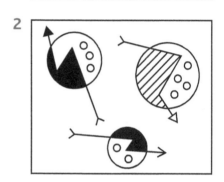

Features in common:

_____

_____

_____

_____

Another member of the group:

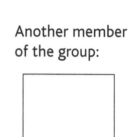

In the next two questions, list the features that occur in some but not all of the members of the group. These are called distractors.

3

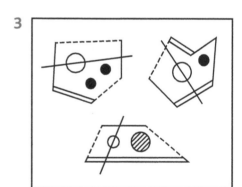

Distractors (features in some but not all):

_____

_____

_____

_____

4

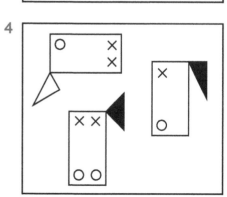

Distractors (features in some but not all):

_____

_____

_____

_____

Test yourself

Look at the first three pictures and decide what they have in common. Then select the option from the five on the right that belongs to the same set. Circle the letter beneath the correct answer. For example:

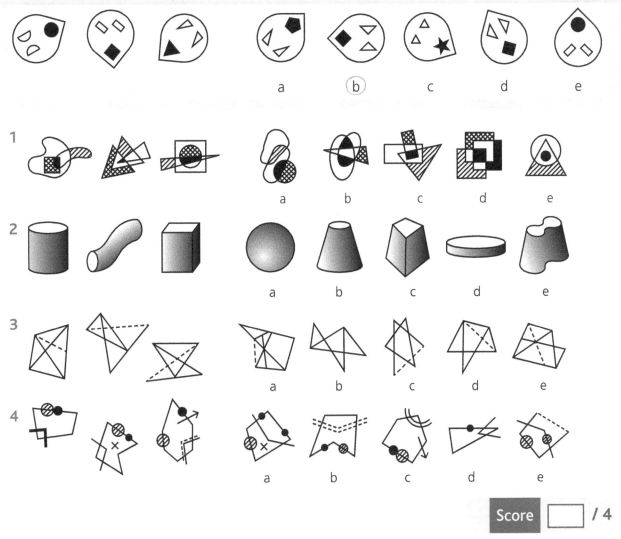

Score ☐ / 4

Try it out

Complete the three patterns on the left so that they are all a bit different from each other, and so that the first pattern on the right belongs to the group and the second one doesn't.

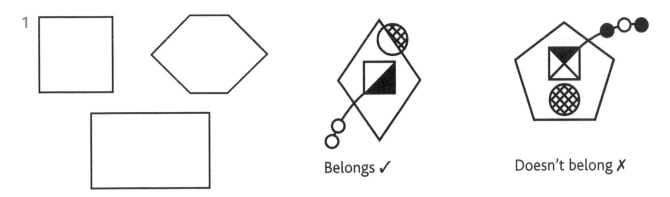

Belongs ✔   Doesn't belong ✗

Applying changes 2

Notice how the pattern changes from the first picture to the second picture in the two pairs. Write the two rules that these pattern changes follow.

1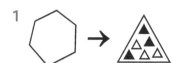

(a) Rule 1: _____

(b) Rule 2: _____

2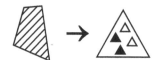

(a) Rule 1: _____

(b) Rule 2: _____

In the next set of questions work out the rules and then apply them to complete the third pair. Draw your answer in the empty box provided.

3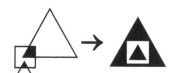

4

5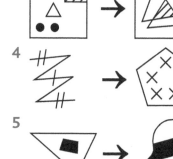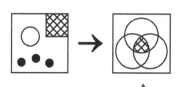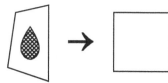

Look at the two sets of pictures on the left connected by arrows and decide how the pictures before the arrows have been changed to create the pictures after the arrows. Now apply the same rule to the third picture and circle the letter beneath the correct answer. For example:

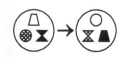

 a   b   ⓒ   d   e

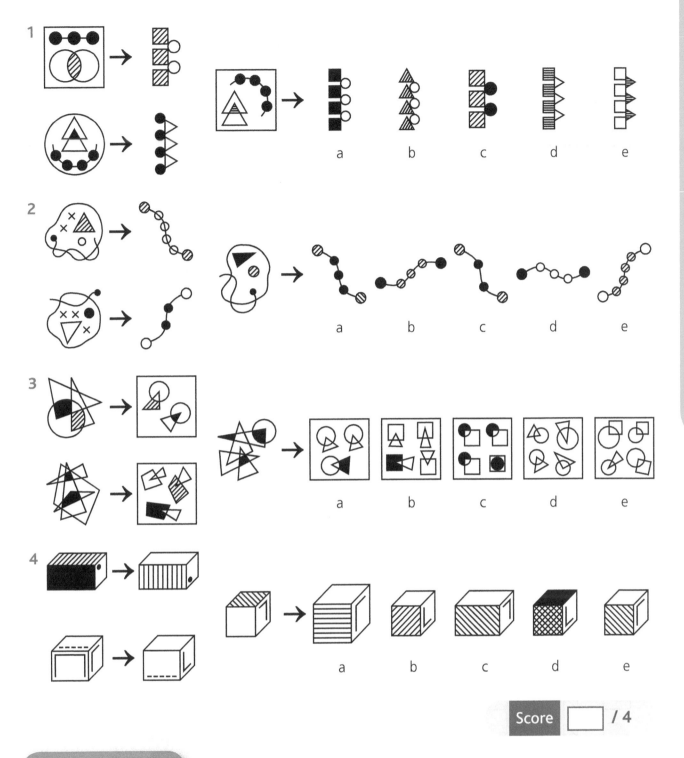

Score [ ] / 4

Try it out

Just one pair of patterns in this example is complete. Draw patterns to complete the second and third pairs so that they have been changed in the same way.

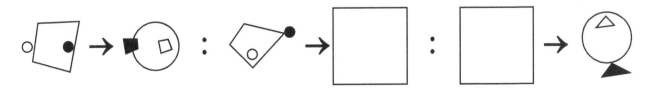

Matching 2D and 3D shapes 2

What flat 2D shape would be seen viewing these stacks of blocks from above? Draw your answer in the space provided.

1 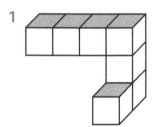 2 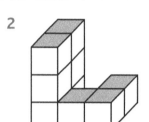 3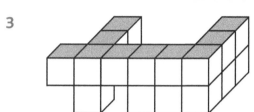

Circle the letters of the **two** 3D shapes that would give the plan drawn if viewed from above.

4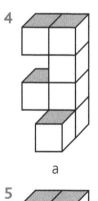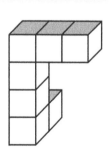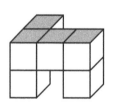

   a   b   c

5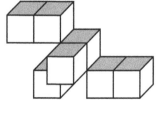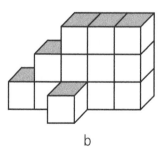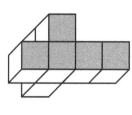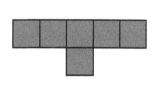

   a   b   c

Test yourself

The answer options are all suggested 2D plans of the 3D picture on the left, when viewed from above. Circle the letter beneath the correct 2D plan. For example:

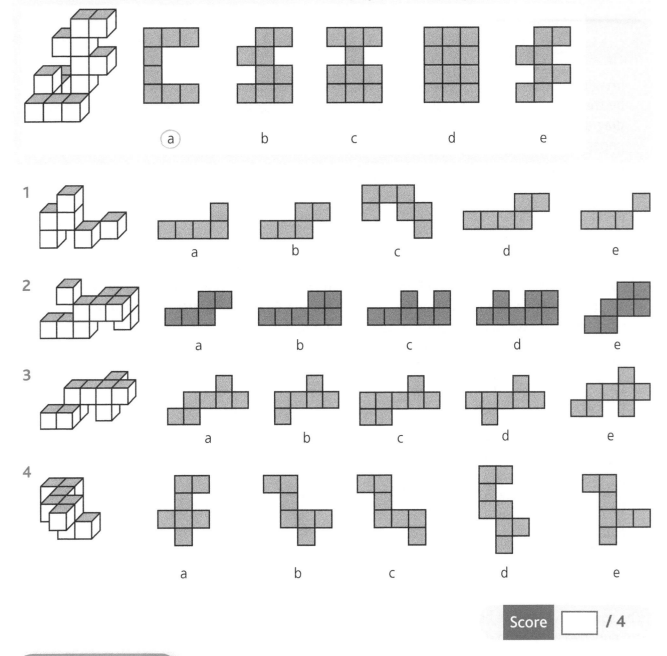

a b c d e

1

a b c d e

2

a b c d e

3

a b c d e

4

a b c d e

Score ☐ / 4

Try it out

How many solid 3D shapes can you make using 12 cubes that would have this 2D plan? If you find drawing 3D cubes hard, use bricks or dice to solve the problem.

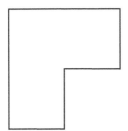

B Position and direction

Following the folds

1 Imagine a square piece of paper being folded along a dashed line in the direction shown by the arrows from the first diagram to the second diagram. Draw shapes on the second diagram where holes would need to be cut to give the pattern in the third diagram after the paper has been unfolded.

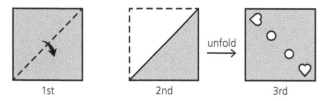

Now the piece of paper has been folded twice, in the direction shown by the arrows. Look carefully at the position of the folds in these diagrams. Draw where the holes will appear on the final diagram when the paper has been unfolded.

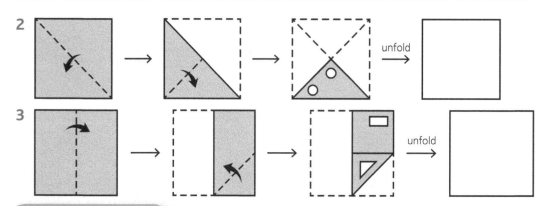

The square given at the beginning is folded in the way indicated by the arrows, and then holes are punched where shown on the final diagram. Identify the answer option which shows what the square would look when it is unfolded. Circle the letter beneath the correct answer. For example:

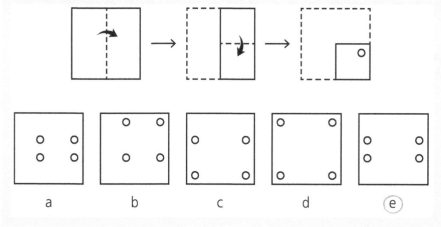

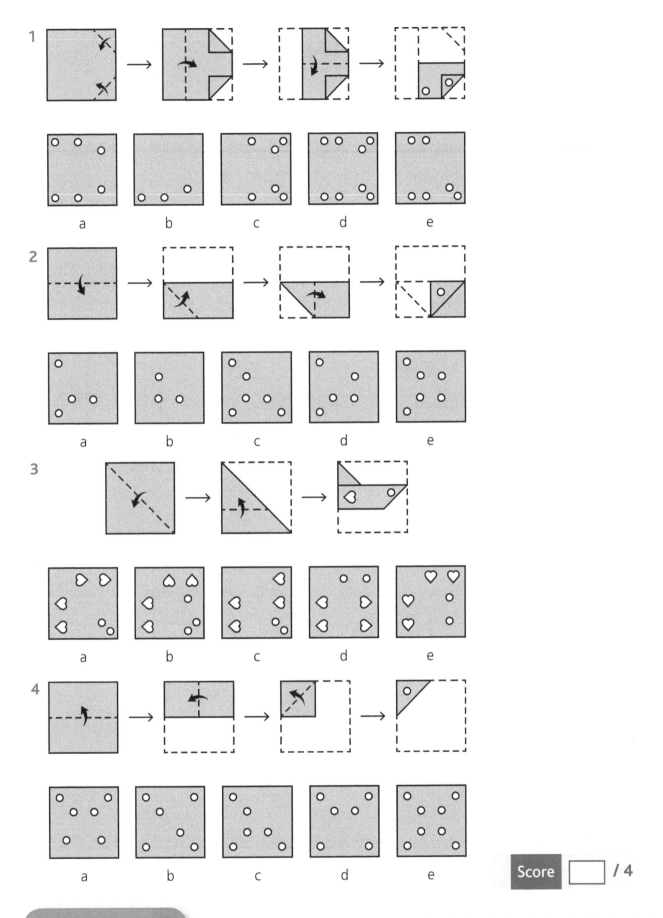

Try it out

Fold a square of paper twice, then punch two holes. Draw the pattern of holes seen when the paper is unfolded. Get a friend or parent to work out how it must have been folded.

Score ☐ / 4

Matching a single image 1

Imagine these sets of blocks have been rotated through space in some way. Which of the three options on the right is a rotation of the set of blocks on the left?

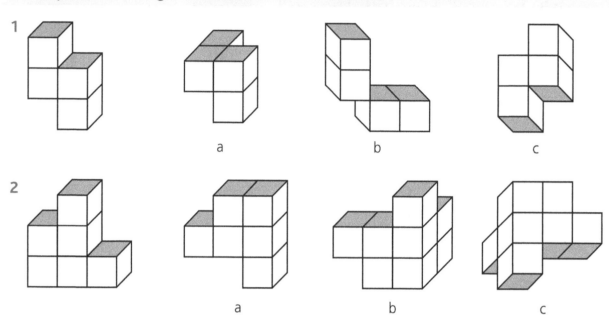

In the next two questions, which set of blocks is different from the other two? Circle the letter beneath the correct answer.

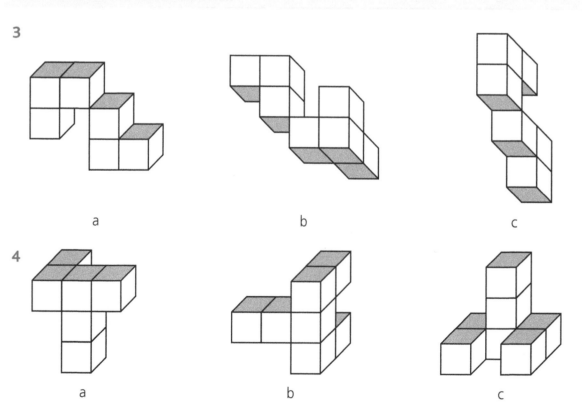

Test yourself

In Questions 1–5 you will see a rotated version of one of the 3D pictures shown here (a, b, c, d, e). Circle the letter of the answer option that indicates the matching shape.

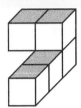

a

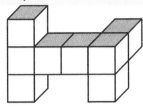

b

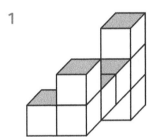

c

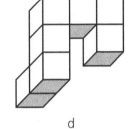

d

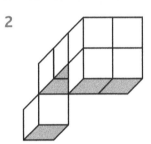

e

1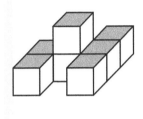

a
b
c
d
e

2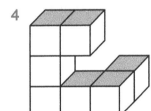

a
b
c
d
e

3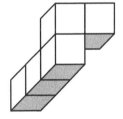

a
b
c
d
e

4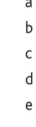

a
b
c
d
e

5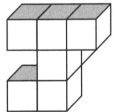

a
b
c
d
e

Score [   ] / 5

Try it out

Using squared paper, try drawing these images viewed from a different angle.

1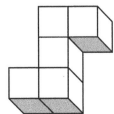

2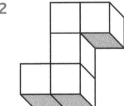

Translating and combining images 1

Look at the small shape. How many can you find in the picture next to it?

1
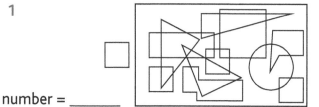

number = _____

2

number = _____

Which two of the shapes on the right **cannot** be found in the picture on the left?

3
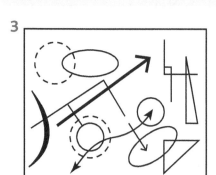

a

b

c

d

e

f

There are two identical shapes hidden within each of the next two pictures. Draw each shape.

4
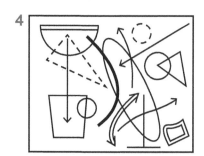

5
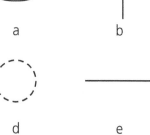

Two of the shapes on the right are hidden in the picture on the left. Circle the two letters beneath the answer options that appear in the picture. For example:

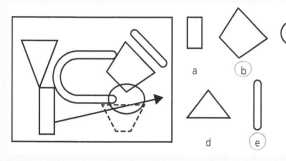

a

b

c

d

e

Score ☐ / 4

Try it out

Have fun drawing a diagram and hiding a shape in it. Get a friend or parent to find the hidden shape.

Matching a single image 2

Look at how the first cube has been rotated so that it ends up in the position shown by the second cube. Rotate the first cube of the second pair in the same way and then draw the pattern that would appear on the **front** face of the last cube.

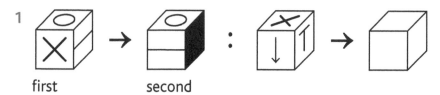

1

first   second

Rotate the second cube in each pair in the same way as the first cube and draw in the patterns on the blank faces. All the cubes have three faces with arrows and three black faces.

2

3

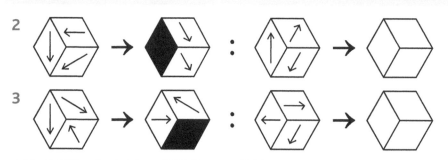

Circle the letter beneath the cube that **cannot** be the **same** as the others.

4

a   b   c   d

5

a   b   c   d

6

a   b   c   d

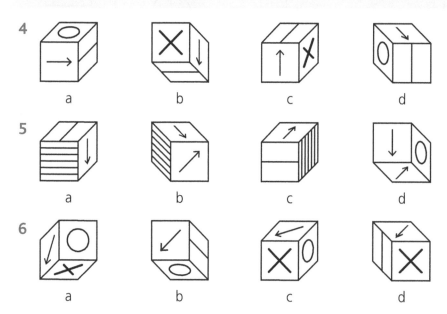

Answers

Please note that all questions are worth one mark unless stated otherwise in brackets.

Most unlike (page 8)

Have a go

1 d **shape – d** is five-sided; the others have six sides
In common: irregular polygons, dashed straight line passing across the shape with three different style lines at the ends

2 d **angle** – there is no right angle along the three-sided angled line in **d**
In common: all have a circle with a three-lined zig-zag crossing the circle; all but **d** include a right angle

3 b **line style** – all but **b** have a circle crossing a dashed line on the quadrilateral and two dashed sides
In common: all have a circle, a quadrilateral with two dashed lines and two plain lines and a shaded triangle inside the quadrilateral

4 d **line style** – all but **d** have a right-angled zig-zag made up of both double and single lines; **d** has only double lines
In common: all have a right-angled zig-zag crossing a pentagon, and all have two black circles side by side on one of the lines

5 b **shape – b** has two triangles; the others have a triangle and a quadrilateral. The new figure should have a triangle and a quadrilateral with two sets of angled lines. One line of each pair should be dashed. The point of the angles should be somewhere within the triangle

6 c **number** – there are six circles in **c**; the others have five. The new figure should have five circles. There does not have to be a pattern to the overlapping circles, which should include a black spot and a square

Test yourself

1 a **shape – a** is the only figure with no cube
In common/distractors: **shading** – on the shapes is random; **rotation** – of the figures is random

2 b **number** – all have five lines except for **b**
In common/distractors: **line style** – number of dashed lines varies; **shading** – on the shapes is random; **shape** – (a) shapes made by the lines are random, (b) all have two squares/rectangles in the shape; **position** – of rectangles is random

3 c **number – c** has ten circles; the others have nine
In common/distractors: **line style** – all wavy; **number** – (a) the number of lines is always the same, (b) the number of arrowheads is random

4 e **angle** – all the figures include one right angle except for **e**
In common/distractors: **line style** – (a) all the figures have one thick line, (b) all have two dashed lines, (c) number of thin lines varies; **shape** – made by the lines is random; **position** – the lines within the figures are random

Try it out

There should be patterns in each box, with some features common to all and one or more features different in a single box, so that it is the odd one out.

Matching features 1 (page 10)

Have a go

1 In common: two overlapping circles, one part of smaller circle shaded, a cord across larger circle, smaller segment of large circle shaded, crosses inside the larger segment
Distractors: number of crosses, style of shading, relative position of elements

2 In common: a triangle with three thin lines, a second dashed-line triangle drawn onto one of the sides, small white circle divided into four in solid-line triangles, small black circle in dashed-line triangle
Distractors: size of black circles, shape of triangles

3 Differences: angle between the two long lines, angle of short lines across main L-shaped line, number of short lines across long line
Distractors: black circles and arrowheads

4 Differences: shading of concentric circles, shading of circles at end of wavy lines, shape of polygon, relative position of circles at end of wavy lines
Distractors: number of intersections

5 Must be a square, any two square corners marked and a half-shaded circle in centre

6 Must have two overlapping triangles, small triangles drawn inside lower section of both triangles below the intersecting part, small circles in the top section of the overlapping triangles
Distractors: shading of small shapes, type of triangle

Test yourself

1 d **number** – has two zig-zag lines, each with three sections, three crossing points; **shape** – same shape at all four ends of zig-zag lines

Distractors: **shapes** – the arrows and circles at ends of the lines, and the enclosed shapes are quadrilaterals; **shading** – whether these shapes are black or white

2 c **shape** – (a) square inside a polygon, (b) same polygon inside half of square; **proportion** – other half of square divided into two parts; **shading** – one-quarter of the square
Distractors: **shading** – style is random

3 e **shape** – zig-zag lines, curved lines; **line style** – double lines for zig-zags, thin and dashed double line for curved lines; **number** – all but one of the zig-zag sections crossed by separate curved lines; **shape** – small circle at one end of curved line
Distractors: **angles** – the zig-zag angles are random; **number** – the number of sections in zig-zag lines is random

4 e **shape** – (a) quadrilateral divided into two triangles, (b) small circle inside one triangle, small triangle inside the other large triangle; **proportion** – small triangle divided into two parts
Distractors: **shading** – in small triangle, of small circle; **shape** – type of quadrilateral

Try it out

Both rectangles must have elements in common with the target shape chosen.

Applying changes 1 (page 12)

Have a go

1 Must be a trapezium with top short and left diagonal lines dashed, contain a large circle with vertical-line shading, double solid thin line arc inside along lower edge

2 Must be a hexagon with an unshaded rhombus inside; the rhombus contains two crosses

3 Must be a pentagon with five horizontal lines across it and four diagonal lines across

4 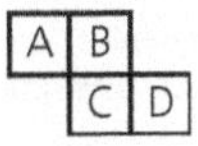 A – square cross-hatch shading; B and D – diagonal shading top left to bottom right; C – diagonal cross-hatch shading

5 First square with diagonal shading bottom left to top right; second square divided into quarters diagonally with left and right quarters shaded black; third square with horizontal-line shading across square

6 One circle white, two black, three with cross-hatching, four with diagonal shading bottom left to top right

Test yourself

1 d **shape** – rectangle divided into three; **shading** – (a) outer sections shaded, (b) shading style diagonal cross-hatch lines, as in the original figure; **number** – the line on each corner has the same number of lines as in the original figure, i.e. two

2 b **number** – four horizontal lines, one vertical line; **line style** – horizontal lines wavy, vertical line straight (the feature of the vertical and second horizontal lines changes position); **shape** – (a) solid arrowheads at both ends of horizontal lines, (b) circle at each end of vertical line; **shading** – black circles at both ends of straight line
Distractors: number, style and position of arrowheads on horizontal lines

3 d **shape** – rhombus containing two circles and a triangle; **line style** – dashed line around the rhombus; **shading** – internal shapes all shaded black
Distractors: size of circles

4 e **number** – eight cubes; **shape** – regular cuboid; **angle** – cuboid viewed from lower right

Try it out

Many answers possible – one suggestion for each given:

1 First pattern: same triangle and quadrilateral overlapping, with area excluding the overlap shaded. Second pattern: an overlapping circle and pentagon with black shading of overlap. Third pattern: same overlapping circle and pentagon with area excluding the overlap shaded.

2 First pattern: same shape with arrow pointing down and black and white circles exchanging place. Second pattern: L-shape with white circle on top of upright and black at end of horizontal line, and horizontal arrow pointing left. Third pattern: L-shape with black circle on top, white at end of horizontal line and horizontal arrow pointing right.

Matching 2D and 3D shapes 1 (page 14)

Have a go

1 (a) e (b) f (c) b

2 c – net must be made up of six equilateral triangles only

3 Four horizontal lines across the face

4 White square with horizontal lines across the square

5 Heart on its side with the point of the heart on the right next to the face with the white circle with the horizontal solid line

6 Right-angled triangle with the right-angled corner being in the top-left corner of the face as it is drawn

Test yourself

1 a When the net is folded the open side of the C-shape will be next to the circle. The side with the X folds down next to the circle and is on the same side as the C.

2 d When the net is folded the X with one part shaded will be opposite the three white circles, the black spot will be opposite the white circle and the U-shape will be opposite the X shape with no shading, so none of these pairs of patterns can appear next to each other on the cube; then check the position of the shading in the X-shape with part shading – the part shaded is along the side away from the black spot, so the X-shape must have its black triangle along the side next to the white circle.

3 d When the net is folded the open side of the two C-shapes must be next to the open side of the V or the X shape, the curved side of the C-shapes can be back to back. The side of a C-shape cannot be next to the X, and if the two V shapes are adjacent they will be pointing in opposite directions.

4 e When the net is folded the two octagons will be opposite each other and the square faces around the sides must appear in the same order as in the net. Look carefully at the directions of the arrows and note which part of the top or bottom shape they point to. The white triangle and the U-shape also have a specific orientation with respect to the top and bottom – the base of the U-shape and the top of the equilateral triangle must be next to the face with the black and white circle.

Try it out

Copy the net you have drawn onto squared paper and cut it out, fold to make cubes and then check your answer.

Matching features 2 (page 16)

Have a go

1 **shape** – (a) all large shapes are quadrilaterals, (b) all contain a small 2D shape; **line style** – (a) three straight and one curved line in each quadrilateral, (b) a curved arrow present; **shading** – all polygons are white

2 **shape** – (a) large circle, (b) small white circles; **line style** – (a) zig-zag solid line, (b) extending across circle line at both ends; **angle** – two angles along zig-zag line; **shading** – (a) one part of circle shaded, (b) small white circles in unshaded part; **proportion** – tail end of zig-zag arrow extending twice as far outside the circle as the arrowhead end

3 **number** – (a) of circles, (b) of sides of the polygon, (c) of each type of line style; **shading** – of circles; **line style** – the order of each line style around the polygon

4 **shape** – type of triangle; **number** – (a) of crosses, (b) of small white circles; **shading** – of triangles

Test yourself

1 c **number** – each figure has three overlapping shapes; **shading** – (a) black where all three overlap, (b) two other elements are also shaded
Distractors: **shape** – of polygons; **shading** – style of other sections

2 d **shape** – each figure is a polyhedron with shape of cross-section constant throughout
Distractors: **shape** – of faces, flat or curved faces; **angle** – between faces

3 d **shape** – each has overlapping triangles; **line style** – one line dashed, the rest solid; **number** – (a) five triangles formed, (b) one quadrilateral
Distractors: **shape** – type and size of triangles

4 d **line style** – solid line round polygon; **shape** – (a) two circles on the sides of the polygon, (b) a V-shape taking in one angle of the polygon, (c) a quadrilateral formed between the V-shape and polygon; **number** – two circles; **shading** – one circle black and one with diagonal line shading
Distractors: **line style** – of V-shape; **number** – of sides of the polygon

Try it out

1 Circles to be crossed by a wavy line with a plain arrowhead, two crosses in one part of the circle and one black circle in the other part

2 A square, half shaded, inside each polygon with a curved line from one corner extending across the polygon and with two white circles on the tail outside the polygon

Applying changes 2 (page 18)

Have a go

1 Rules: (a) number of sides of polygon gives number of small triangles inside a large triangle, (b) half of the small triangles shaded

2 Rules: (a) all three shapes drawn inside each other, (b) the shading pattern is used for inner and outer shape

number – two of the same shape and one other; **shading** – for inner and outer parts of new shape; **proportion** – of shapes relative to each other with smallest inside
Distractor: **position** – of shape in first diagram

3 Must be two overlapping circles with overlapping section shaded black
number – of black spots gives number of shapes; **shape** – given by the white shape in the square; **shading** – given by the shading of the top right corner

4 Must be a quadrilateral with three crosses inside
number – (a) of lines in zig-zag gives number of sides of polygon, (b) short lines across zig-zag gives number of crosses inside shape
Distractors: **size** and **angle** of the short lines

5 Must be an irregular shape with four lines going across the shape and alternate bands formed with cross-hatch shading
number – sides of polygon give number of lines; **shading** – of circle gives shading style for bands
Distractors: **shape** – of shaded part inside the polygon; **proportion** – widths of the bands across the irregular shape

Test yourself

1 d **shape** – (a) large outer shape of each pair gives the shapes for the first column, (b) the overlapping shapes inside give the shape for second column; **number** – of black circles gives the number of shapes in first column; **shading** – of the overlapping shapes inside the first shape gives shading style for first column shapes
Distractors: **proportion** – of the elements within first shape

2 b **number** – of intersections of curvy line across shape outline gives number of beads in centre of string; **shading** – (a) of small circle in shape gives shading for beads, (b) of triangle gives shading for circles at each end of the 'string'
Distractors: **proportion** – of the shapes in the first pattern

3 d **number** – of triangles in the first shape gives the number of elements inside the square; **shape** – the shape overlapping with the triangles in the first figure is the shape that will overlap with the triangles inside the square in the second figure
Distractors: **shading** – number of shaded sections

4 e **size** – the second cuboid is the same size as the first one in each pair; **shading** – shading or pattern on top face changes to front face
Distractors: **size** – dimensions of cuboid

Try it out

Many variations possible. All should have the **shape** of the large polygon becoming the shape of the two smaller shapes, one of which will be inside and one outside a larger circle; **shading** – of the small inside shape in first pattern becomes shading of the small outside shape in second pattern and vice versa.

Matching 2D and 3D shapes 2 (page 20)

Have a go

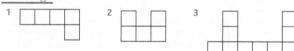

4 b and c   5 a and b

Test yourself

1 e On the front row there are two cubes on the bottom layer of the stack, the gap between them is covered by one of the two cubes of the second layer, and cubes in the third layer cover the end and central position, so from above this is a row of three. There is no cube in the corner but on the bottom layer is one cube going back as is from the corner position, so the plan will be an L-shape one cube wide throughout with the corner square missing.

2 b The front row has three cubes on the bottom layer and in the second layer one cube covers the end cube and then extends for another two places to the right. The single cube in the third layer is over the second cube on the first layer, so the overall plan for the front row is a line of five squares. There are more cubes forming a second row behind the two cubes at the right-hand end, so the second row is just two squares long from the right, giving a plan of three in a row joining onto a 2 × 2 square.

3 a At the left end on the bottom layer there are three cubes in a backward L-shape, above the back right cube is a cube on the second layer, which extends for four cubes to the right. Counting along this row of four, there is one cube extending backwards from the third cube all on the second layer, and a single cube beneath the third cube on the bottom layer.

4 b Starting with the top (second) layer at the back, there are two cubes next to each other, with two more coming forward from the right-hand one, then one cube extending to the right of the third cube and an extra one extending forward off the end cube. On the bottom layer beneath this there is an additional cube extending to the right giving a plan that is a three-long, three-wide L-shape, with a square on the left at the top of the L and a square off the middle of the base line of the L-shape.

Answers will vary.

B Position and direction

Following the folds (page 22)

Have a go

1   Vertically orientated heart-shaped hole in the lower right corner of the triangle, and one circle hole midway along the hypotenuse inside the triangle
2   Two lines of four circles in a diagonal line, from lower left to upper right across the square
3   Rectangle going across the top-left and top-right corners, four right-angled triangles in the lower half, the two on the right being a mirror image of the two on the left, which have the right-angled corners in the bottom-left and top-right corner of the quadrant, with a small gap between the two hypotenuse lines

Test yourself

1   d   With each unfold, the fold line becomes a line of symmetry, so going back one stage there will be a circle at the top and the bottom near the fold line and one by the diagonal line also at the top and one at the bottom; then undoing the next fold will add another circle at the top and another at the bottom of the left-hand side of the square; and undoing the last two diagonal folds gives a second circle in each corner on the right.
2   e   Undoing the last fold gives two circles, in second and third place on the third row, with the vertical halfway line a mirror line; unfolding the triangle on the lower left gives a third circle in lower left corner; then unfolding top half from lower half gives a mirror image of these three circles in the top half, as in option **e**.
3   b   Divide the square into a 3 × 3 grid; then unfolding the trapezium downwards will give two heart shapes, both on their sides with the point on the left, one in the first square of the middle row and one in the first square of the third row, and one circle in the triangle at the end of the third row; then unfold the diagonal triangle, which will give the diagonal fold line as a line of symmetry or mirror line, resulting in two hearts, one in the second and one in the third square along the top row, each with their points at the top, two hearts on the left as before, and two circles in the third square of the third row and one circle in the third square of the second row as in option **b**.
4   e   Unfolding the first fold gives two circles in top left to lower right diagonal line in the top quarter, unfolding the next fold gives a mirror image of that pattern in the top-right quarter, and when unfolding the last fold all four circles along the top half are reflected in the lower half, giving eight circles making two diagonal lines right across the square as in option **e**.

Try it out

Many answers possible.

Matching a single image 1 (page 24)

Have a go

1   c   rotating solid 180° forward gives option **c**
2   a   rotating solid upside down 180°, then clockwise 90° gives option **a**
3   b   only **b** has cubes going in two planes
4   a   the L-shape formed by cubes at one end of this solid is made up of three cubes in **b** and **c** but four cubes in option **a**

Test yourself

1   b   shape **b** is rotated 90° clockwise around a vertical line and rotated 180° round a horizontal line
2   e   shape **e** is rotated 90° clockwise round a vertical axis
3   a   shape **a** is rotated 180° round a vertical axis
4   d   shape **d** is rotated 90° anticlockwise round a horizontal axis
5   c   shape **c** is rotated 90° anticlockwise round a horizontal axis and 90° clockwise round a vertical axis

Try it out

Answers will vary.

Translating and combining images 1 (page 26)

Have a go

1   3: bottom left, middle and bottom right
2   2: one as a U-shape in centre along top edge and the other as a C-shape in lower right corner
3   b and e
4   one on the left pointing down, the second pointing down diagonally from top-left corner towards the centre
5   one at the top of the lower left quadrant, the second in the bottom-right corner

Test yourself

1   c and e   **a** is too wide, there is no trapezium shape like **b** in the pattern, the Z-shape in **d** has different proportions from the one in the diagram, but **c** is in the lower left corner and **e** is in the centre of the bottom half
2   a and e   **b** is a pentagon and there are no pentagons in the pattern, there are no V-shapes with an acute angle and wide lines so not **c**, and **d** does not appear; **a** is rotated in the lower left quadrant and **e** is in the top right part of the lower right quadrant
3   b and d   not **a** as the double-lined C-shape is wider in the diagram, there is no plain white circle **c**, and no straight simple arrow **e**; **b** in the middle of the left half on the top and **d** in the lower right corner
4   a and b   there are no straight lines with a short angled line off the end so not **c**, no E-signs with the double line so not **d**, and no right-angled Z-shape ending with a small T-piece; **a** is in the middle of the top-left quadrant and **b** appears rotated in the lower right corner

Try it out

Answers will vary.

Matching a single image 2 (page 28)

Have a go

1   A vertical T-shape the right-hand side face comes to the front as the cube is rotated 90° clockwise.
2   The top right face is rotated 90° anticlockwise, so the arrow on the top face will point to the top-left corner, the face on the left becomes the face at the front lower right with the diagonal arrow going up from bottom right to top left and the left face of the cube is shaded.
3   The lower right face rotates up 90° giving an arrow pointing from the top edge straight to the middle of the bottom edge on the top-right face, the arrow on the left face also rotates 90° anticlockwise, and will end pointing diagonally across the face, straight down towards the bottom corner on the diagram, and the front lower right face is shaded.
4   d   the circle is on the right of the downward pointing arrow and the cross on the left of it in the other three cubes
5   b   the line shading on the front left face is not perpendicular to the straight arrow on the top face
6   b   it is not possible for the straight arrow to point to the face with the circle

Test yourself

1   d   The face with the downward pointing arrow has been rotated to the right so the line on the front face is now a vertical line rather than a horizontal line, so the cube has been rotated 90° anticlockwise. Applying this to the second cube will give the arrow pointing to the left across the right-hand face and the line across the lower front face will become a horizontal line.
2   c   The front face rotates 90° upwards so the line on the left side face rotates 90° backwards as well. Applying this to the next cube means that the front face will go to the top – an arrow pointing to the right, and the left-side face will rotate 90° backward leaving the arrow pointing forward and slightly down, as in **c**.
3   e   The front face rotates 90° anticlockwise so the pattern on the top face will rotate and appear on the left-side face; applying this the single line on the top will rotate onto the left-side face running from back to front and the lower right face will show a line from bottom to top.
4   d   Left side rotates 90° clockwise to top position and front face will rotate 90° clockwise as well; applying this gives a single line going across the top-right face and two lines from bottom to top on the lower front face.

Try it out

Answers will vary.

Translating and combining images 2 (page 30)

Have a go

1   Overlay dots onto grid and move along four and down one

2   Overlay dots onto grid and move along two and down one

3   Overlay dots onto grid and (1) move down one for the first and (2) move across five for the second

4 Overlay dots onto grid – that gives the first, then move along five and down one for the second

5 Rotate 90° anticlockwise then overlay onto grid and move along two

6 Rotate 90° anticlockwise and overlay onto grid

Test yourself

1 The pattern is revealed by overlaying the pattern of dots in the first box onto the second box and moving it along one place to the right and down one place.

2 The pattern is revealed by overlaying the pattern of dots in the first box onto the second box and moving it along one place to the right and down two places.

3 The pattern is revealed by rotating the shape 90° anticlockwise, then overlaying the pattern of dots in the first box onto the second box and moving it along one place and down one place.

4 The pattern is revealed by rotating the pattern of dots 90° clockwise then overlaying the pattern of dots in the first box onto the second box and (1) moving along two places and down one and (2) along four places and down one.

Try it out

Answers will vary.

Translating and combining images 3 (page 32)

Have a go

1 b is made up of six cubes only
2 b is made up of nine cubes only
3 a, c and e the top cube e sits on the corner of an equal-armed L-shape; c, which in turn has its corner resting on one half of a 2 × 1 × 1 block, a
4 c, d and e the top regular T-shape c is resting on a 2 × 1 × 1 block; d, resting on a single cube, e
5 a, d and e the U-shape d fits around one arm of the L-shape a and has one corner resting on the top end of a vertically placed 2 × 1 × 1 block, e

Test yourself

1 c the end of a 3 × 1 × 1 block fits into the bottom part of the T-shape, which rests on a 2 × 1 × 1 block
2 c a 2 × 1 × 1 block rests along the top edge of an upside-down back-to-front L-shape and a second L-shape sits over the end of the 2 × 1 × 1, at right angles to the first L-shaped block
3 a there are three 2 × 1 × 1 blocks, two lying flat forming an L-shape with the third standing vertically with its base in the corner of the L-shape, and a single cube sitting outside the L-shape on the top-left side
4 d the base of a T-shape rests on one end of a 2 × 1 × 1 block with a 3 × 1 × 1 block standing vertically behind the right-hand arm of the T-block and a single cube resting centrally on the top of the T-block
5 b a 2 × 1 × 1 block is standing vertically at the back next to the vertical arm of an L-shape, with a single cube in front of it, and a second single cube to the left, adjacent to the first cube, extending away from the L-shape

Try it out

The following diagrams represent a few of the many possible answers:

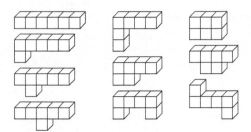

Translating and combining images 4 (page 34)

Have a go

1 a 3 b 2 c 1 d 4
looking at the thick line and lining up the squares so that the thick line forms a wide upward curve along the top (iii next to ii) and a narrower downward curve in the lower half (i next to iv)

2 a 5 b 2 c 6 d 4 e 1 f 3
looking at the horizontal line just over half way down, it goes right across and at the right-hand end has a diagonal line coming up to join it at the edge; this gives up pieces iv, i and iii in that order in positions d, e and f. Along the top row the dashed line is in the top part of the square on the left and the lower part on the right, and two dashed lines cross over in the middle, giving us pieces v, ii and vi along the top row in places a, b and c

3

e	i	b	k
f	h	g	a
l	j	c	d

The top-right square (k) will have a horizontal arrow pointing into the square from the left edge about a third of the way down the side.
The bottom-left square (l) will have a solid diagonal line from bottom-left corner to top right and a thick line starting one-third along the bottom very gently beginning to curve up, coming to the right-hand edge of the square about a third of the way up.

Test yourself

To complete these jigsaws work systematically! Visualise the complete pattern divided into squares, identify the features in the first square and find those features in an answer option. Write the letter in its place on the grid. It may help to draw pencil lines over the pattern dividing it into the correct number of sections. Distinctive features can often be identified straight away, such as the thick black curved line in the first picture, but other parts need to be checked more carefully.

1

l	y	h	c	dd
j	cc	f	e	p
aa	u	r	b	w

z	o	a	t	i
g	d	n	q	v
bb	x	k	m	s

2

m	h	e
b	p	a
q	k	r

g	l	i
f	d	n
j	c	o

To solve these puzzles identify specific distinctive features in the grid, such as thick lines, distinct shapes or shading, arrowheads or dots. Notice where they are located in the grid and then look systematically through the squares to find the one that includes that feature and record its number or letter in the blank grid.

Try it out

There should the same number of squares as sections in the grid. Double check the accuracy of your copying by testing the exercise out on a friend or parent.

Maths workout 1 (page 36)

Building 2D and 3D shapes

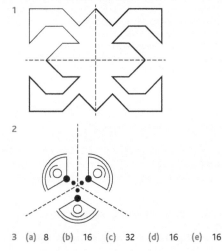

3 (a) 8 (b) 16 (c) 32 (d) 16 (e) 16 (5 marks)

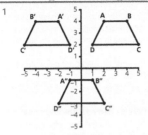

2   Answers to (b) and (c) are shown on the diagram in question 4.
    (a)   A' (−2,4)   B' (−4,4)   C' (−5,2)   D' (−1,2)
          This image is a reflection so remember the labelling
          needs to show that.
    (d)   A'' (−1,−1)   B'' (1,−1)   C'' (2,−3)   D'' (−2,−3)
          Remember that the *x* axis value always comes first.   (4 marks)
                                                                  (3 marks)
3

C Codes, sequences and matrices

Connections with codes 1 (page 38)

Have a go

1   X, L, NX, Y, M
    shape – L is for square, M for triangle and N for circle; **number** – X is for five
    circles, Y for four and Z for three
2   X, BY, A, Z, A
    angle – A is for oblique lines, B for vertical and C for horizontal; **position** – X is
    for outside, Y is for on the line and Z is for inside
3   BN: **shape** – A is large oblong, B is small oblong and C is square; **number** – L is
    for three lines, M is for four and N is for five
    Distractor: **position** – the placement of the lines is not relevant
4   CLY: **shape** – A is for shaded circle, B for shaded triangle and C for shaded
    diamond shape; **number** – a) L is for two white circles, M is for three and N is
    for four, b) X is for four short curly lines, Y is for three and Z is for two
    Distractors: **shape** – of the triangles is irrelevant; **position** – the orientation of
    the complete figure is irrelevant
5   EZC: **line style** – D is for double solid line, E for one dashed and one solid line,
    and F is for two dashed lines; **angle** – X is for horizontal, Y is for diagonal and Z
    is for vertical; **number** – A is for one circle, B is for two and C is for three
    Distractors: **shading** – of circles; **position** – of circles against the lines

Test yourself

1   e   **shape** – A is for large trapezium, B for large parallelogram; **number** – of
        white circles, X for three, Y for two and Z for one
        Distractors: **shading** – of other polygons; **shape** – of small polygons;
        position – of white circles; **size** – of white circles
2   d   AZN: **position** – of smaller circle, A for inside large circle, B for on the line
        of the larger circle and C for outside larger circle; **line style** – X is for short
        curved line, Y for V-shaped line and so straight line will be Z; **shading** – M
        is for line shading and N is for black shading
        Distractors: **position** – of the lines crossing the circle, **line style** – of large
        circles
3   a   BZL: **shape** – (a) A is for circle inside shape, B is for oval inside shape; (b) X
        is for square outline shape, Y for Z-shape and Z for outer L-shape; **shading**
        – L is for black shading and M is for lined shading
        Distractors: **orientation** – of the outer L or Z shape; **position** – of smaller
        shape within the larger shape
4   c   EMY: **shading** – D for black circle, E for white circle and F for double circle;
        number – L is for five lines in the zig-zag, N is for six lines so M is for four
        lines; **line style** – X is for solid line, Y is for dashed line and Z is for thick
        line
        Distractors: **angle** – of the lines in the zig-zags; **length** – of elements
        within the zig-zag; **size** – of circle at the end of the zig-zags

Try it out

There should be a pattern in each of the three boxes on the left with common
features linked by code letters. The code for the target figure on the right must be
possible to work out from the codes given on the left.

Sequences 1 (page 40)

Have a go

1   U-shape is rotating 90° anticlockwise so the fourth term should be an inverted
    U-shape and in the fifth term the U-shape will be as in the first term.
    The black and white circles are alternating – so white circle in fourth term and

black in the fifth term. All terms are underlined with a single straight line that
does not touch the U-shape.
2   The number of lines radiating out from the central black spot is increasing by
    one each time, so there will be four lines in the fourth term and five in the
    fifth term. The radiating lines alternate between having a curved end and being
    straight; they will be straight in the fourth term and have curves in the fifth
    term. The first line radiating out in each term has progressed clockwise by one-
    twelfth, or 30°, so the first of the four lines in the fourth term will commence
    straight down from the centre, and the first curved line in the fifth term will
    start 60° before the vertical upright position – pointing to the 10 if the circle
    was a clockface.
3   The L-shape is rotating 90° clockwise between each term. The number of
    the short lines at the end of one arm of the L increases by one each time
    progressing along the sequence, so the first term will have an L-shape with its
    corner in the top right and there will be a small black circle at one end and one
    short line at the other. The fourth term will have the L-shape with its corner in
    the top left, a black spot on one end and four short lines.
4   There is a pattern of three along the top-left corner of each square – white
    circle, cross, black circle.
    The small central L-shape is rotating 90° anticlockwise round the central point.
    The small black spot along the bottom is progressing along the bottom of the
    square, from left to right.
    So the first term or square will have a small black spot in the bottom left
    corner, a small L shape in the centre in the same orientation as the fifth square,
    and there will be a small white circle in the top left.
    The fourth term or square will have a small black spot along the bottom edge
    just before the middle line, a small white circle in the top-left corner and a
    backward L-shape in the centre opening to the top right hand corner of the
    square.
5   (a)   The direction of the central horizontal arrow is alternating pointing left
          and right.
    (b)   The short 'leg' at the bottom of the diagram is rotating 90° clockwise
          each time.
    (c)   The pattern of three should be a set of three circles the same size, one
          with black shading, the next with lined shading and the third one white.

Test yourself

1   d   **position** – (a) the short black line is moving round the edge of the square
        rotating 45° each time, (b) the small black line at the bottom is moving
        across from left to right; **shading** – in the centre of the small square goes
        black, diagonal, cross, then a gap then black again. The diagonal could be
        the same as the second term if it's a repeating sequence of three, or it
        could be in the other if it is a symmetrical sequence – the answer choices
        will indicate which it is
2   d   **number** – (a) the polygon has a decreasing number of sides – nine, eight,
        seven so far, the number of short lines across the base of the arrow is
        increasing by one each time; **shape** – the arrowhead is alternating plain
        line style and white triangle; **shading** – of the circles on the arrow could
        be a repeating pattern or a symmetrical pattern – look at the answer
        options to know which it is
        Distractor: **angles** – in the polygons
3   e   **position** – (a) of the shaded square is rotating round the square clockwise,
        (b) the circles are always in the square section following the shaded
        quarter around the square; **number** – of circles increases by one each
        time; **shading** – (a) of the circles added alternates between adding black
        and white, (b) of the square quarter follows a repeating pattern diagonal
        up, diagonal down, horizontal, vertical, diagonal up, etc.
        Distractor: the arrangement of the circles within the square section
4   b   **size** – plain arrow alternates between long and short; **angle** – (a) the
        angle of the plain arrow moves 45° clockwise between successive terms,
        (b) the angle of the short dashed line moves 45° anticlockwise with each
        term; **shading** – the circle at the end of the short dashed lines alternates
        from black to white
5   e   **number** – (a) number of short lines crossing the curvy line increases by
        one each time, (b) the number of black spots decreases by one each time,
        (c) the number of horizontal lines across the first square decreases by one
        each time; **position** – the position of the horizontal line removed from
        the first square is first the lower of the middle two lines, then the upper
        one, so would then be lower of the outer two lines and then the upper
        Distractor: **position** – of short lines across the curved line

Try it out

All the elements do not need to be used in the sequence as some could be kept as
distractors.

Matrices 1 (page 42)

Have a go

1   A central square with uphill diagonal shading lines, a small white circle in lower
    left corner and a small white circle in top right corner
    Reason: from looking at the pair of squares across the top of the grid you can

see the central shape stays the same, the position of the small circles is the same but their shading changes from black to white

2 An equilateral triangle, with one line at the top of the square parallel to the top edge and with horizontal lined shading
Reason: same shape in each column, shading style moves down one square moving across the grid giving one shape with each shading style in each column

3 A heart shape on its side, with vertical line shading, point towards centre, small arrow in top left corner pointing to the corner, and small arrow in lower right corner pointing diagonally towards the central line
One line of symmetry – vertical line midway across the grid

4 Straight lines turning with right angles making a question-mark shape with arrowhead pointing out to the right, a square in the lower right corner with a diagonal bottom left to top right corner and shading lines perpendicular to the diagonal line across the outer half
Two lines of symmetry – vertical and horizontal lines midway across the grid.

5   6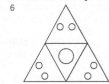

Test yourself

1 b   **shape** – the shape on the left is reflected on the right; **shading** – (a) black shading on the left is white colour on the right, (b) lined shading on the left is black shading on the right, (c) white areas on the left are shaded with horizontal lines on the right
Distractor: the vertical line through the shape and the square

2 a   **shape** – there is a rectangle on the insides of the squares in the middle of each side, surrounding the central square; **line style** – there is a small line in the centre on the outer edge of the middle square on each side; **shading** – there are four styles of shading used: black, white, cross-hatched and diagonal; **position** – the block shading is opposite the cross-hatched shading and the white shading is opposite the diagonal shading

3 d   **line angle** – there is a diagonal line in the top-left square corner of each larger square, alternating from uphill to downhill; **line style** – the short horizontal line has the same style in the squares going diagonally across the larger grid; **shape** – the first column has circles, the second column has small squares and the third column has small triangles; **shading** – the shading of the small shapes is the same along each row – diagonal uphill along the top row, diagonal hatching in middle row and diagonal downhill in the bottom row

4 e   **shape** – the central square is made up of patterns from the top and bottom square, so must contain the inverted V-shape and the inverted L-shape; **position** – the patterns repeated in the central square are in the same relative positions as they are in the square above and below; **translation** – the shapes are just moved across, i.e. there is no rotation or reflection
Distractors: shapes in the other squares, shading, size and position of the other elements

Try it out

Any sensible pattern with one answer option completing the grid correctly.

Connections with codes 2 (page 44)

Have a go

1 The first letter stands for the size of the box: A is large, B is small. The second letter represents the number of black circles: X = 1, Y = 2, Z = 3. Many answers possible for the additional picture, e.g. a large square with one black circle would be AX.

2 The first letter stands for arrow style: D is curved arrow and E is straight arrow. The second letter represents shading style – A for diagonal lines, B for cross-hatch shading; the third letter is for shape – X for a circle, Y for triangle and Z for diamond. Many possible answers for an additional figure, e.g. a circle with cross-hatch shading and a straight arrow cutting across it would be EBX.

3 The first letter could be for the angle of the arrow: A for vertical, B for horizontal and C for diagonal. Many answers possible, e.g. add a circle at the end of the arrow in the first and third figures, so Y represents a circle, add a square at the end of the second arrow, so X is a square and a triangle at the end of the fourth arrow so that Z is for triangle.

4 The first letter will represent the outline shape: L for circle and M for square. Of the many answers possible, the feature coded by the middle letter must be different for each diagram. The feature coded by the last letter must be the same in the second and third diagrams, yet those two different from the first.

Test yourself

1 b   BY: **position** – of the wavy line in relation to the square with A for when the line passes through the square from top to bottom and B when the

line passes through the square from bottom to top; **shape** – ending of the curved line, Y for a white circle, X for an arrowhead and Z for a black circle
Distractors: **orientation** – of the line through the square; **shape** – of the additional part on the side of each square; **shading** – of the additional small shapes; **position** – of the extra shape in relation to which side of the square they are joined to

2 c   AZ: **number** – (a) of circles: A for one, B for two and C for three; (b) of triangles – X for one, Y for two and Z for three
Distractors: **number** – of overlapping sections; **shading** – some sections shaded; **size** – of the circles and triangles

3 d   CGS: **line style** – (a) A for triple solid, B for dashed in the middle and C for solid in between dashed lines, (b) G for straight line, H for curved line; **orientation** – P arrow pointing right, Q arrow left, R arrow down, so need a code for arrow pointing up, so S for pointing up
Distractors: **shading** – of arrowhead; **size** – of triangle of arrowhead; **shape** – at the ends of the arrows

4 b   BNY: **number** – (a) of right angles along bold line, A for three and B for two, (b) number of times the thin line crosses over the thick line – M1, L2 and N3, (c) number of sections in the thin zig-zag line – Y three, X four and Z five
Distractors: **orientation** – of lines; **length** – of elements along each line

5 e   AFZ: **position** – (a) of circle in relation to triangle, A for in the triangle, B across a corner and C on a line, (b) of square – D inside the triangle, E on a line and F outside; **shading** – of square – X is lined, Y is white and Z is black
Distractors: **shading** – of triangle; **angle** – type of triangle; **size** – of circle

Try it out

Make sure there are five diagrams with three code letters written beneath the first four. The first code letter should represent changing shapes, the second code letter should represent different quantities of a single shape and the third code letter should represent some variation in the shading. Check that any extra elements drawn as distractors cannot fit into one of the code patterns.

Sequences 2 (page 46)

Have a go

1 Example of a repeating sequence – large white circle, small diagonally shaded circle (repeating pattern made up of two circles, two squares, two circles, etc.). Example of an alternating sequence – large white triangle, small diagonally shaded triangle (alternating pattern is large version of a shape alternating with a small version of the same shape). There are other possibilities – check carefully and identify either the repeating element or the alternating element

2 Example of a reflective sequence – upside-down L-shape with corner at top left, then a single vertical line (this sequence would have a line of symmetry passing through the middle of the middle figure). Example of a continuing sequence – the next figure would have four lines, all at right angles, with the fourth line added at the top-right end of the U-shape, pointing out to the right, and the next term would repeat that with an additional line at right angles coming down from the end (this sequence would be a continuous sequence, building up a 'battlement' type pattern)

3 • Number of sides of the polygon increasing
  • Number of short lines crossing edge decreasing
  • White circle alternating between being inside and on the line

Test yourself

1 e   **position** – scalloped line progressing down the square; **shading** – (a) block of eight black spots change one at a time to white as progress along, (b) of shape in bottom-left corner alternates black and white; **shape** – in bottom left follows repeating pattern – square, triangle, circle
Distractors: **shape** – sets of small lines; **position** – of spots within the square

2 c   **position** – (a) white square moving round the edge of the square in a clockwise direction, (b) black circle moving anticlockwise around the square; **line style** – small square in the middle with downhill diagonal arrow following an alternating pattern

3 e   **position** – J-shape alternates with the L-shape; **rotation** – J-shape rotates 90° clockwise each time; **shading** – circle in hook of the J changes from black to lines to white

4 d   two alternating sequences: **number** – (a) Z-shape decreasing by one line each turn then increasing again, (b) number of lines on U-shape increasing by one each time; **rotation** – U-shape rotating clockwise 90° each turn; U-sequence is a **continuing sequence**

5 b   **rotation** – the shaded quarter in each square rotates 90° clockwise each time; **shading** – lines alternate from horizontal to vertical; **shape** – the small straight-line pattern alternates with the circle; **number** – (a) small straight line has a decreasing number of cross-lines progressing along the sequence, (b) the white circle has an increasing number of black dots
Distractors: **orientation** – of the small straight line figure; **size** – (a) of the circle, (b) of the spots inside the circle

Seven separate patterns should be shown with the possibility of working out the logical sequence.

Matrices 2 (page 48)

Have a go

1 The square will have a short white-headed arrow in the middle at the top, pointing downwards, there will be a circle with a dot in the top-left corner and a plain arrow pointing left along the bottom on the left side

2 (a) An arrow with a black arrowhead on the inside of the top-left triangle, parallel with the edge of the hexagon

 (b) A short plain line in the top-right triangle near the centre linking the dashed lines cutting across the two adjacent triangles

 (c) A small black circle in the central tip of the lower right triangle

3 (1) circle divided in half with horizontal line and lower half shaded black

 (2) white circle divided in half with horizontal line

 (3) white circle divided in half with horizontal line

 (4) smaller black circle in the centre

4 e one crescent, one black and two white circles (alternate triangles have two or three circles)

Test yourself

1 d **shape** – shaded semi-circle same as in directly opposite triangle; **line style** – (a) V-shape pointing towards the centre as in alternate triangles, (b) short line from top-left corner parallel to the left edge of the triangle – part of a pattern around the edge of the hexagon

2 e **shape** – (a) square, (b) divided into halves vertically; **shading** – (a) background grey, (b) diagonal lined shading – opposite direction of shading in the two halves

3 b **orientation** – arrow pointing left at bottom of triangle; **number** – bottom right corner with three curved lines, part of an increasing pattern going round clockwise; **shading** – central section line shaded to complete central pattern in octagon

4 b **translations** – black spot in top-left corner of top-left square in bottom-right corner of top-right square, and repeated in top-left corner of middle-left square so @ pattern in top-right corner of top-right square repeated in bottom-left corner of top left square and so will be repeated in top-right corner of middle-right square; X-shape in bottom-left corner of bottom-left square repeated in bottom-left corner of middle-left square, so U-shape in bottom-right corner will be repeated in bottom-right corner of middle-right square; short line – to the left of the F-shape in the central square repeated in centre of central-left square, so ^ shape on the right of the F in the central square will be repeated in the centre-right square of the grid

5 c **orientation** – arrow with curly tail pointing up; **shading** – arrows on the left of the grid have single line shading with the full grid shading in the arrow on the right; **reflections** – (a) small L-shaped line – reflected in vertical and horizontal line, (b) half shaded circle reflected in vertical and horizontal lines

6 c **shape** – same shape down each column, so circle; **shading** – same shading along each row – so shading with a ×; small **line pattern** moves down the central line of each square as it progresses from the top row to the bottom, so > shape in centre along lower edge; **number** – of dashes in the short vertical line in each square repeating 1 – 2 – 3 pattern so need one vertical dash; **position** – (a) of small shape – always in top-left corner, (b) of small line figure top centre along top row, middle centre along middle row and lower centre along bottom row, (c) dashed vertical line – starts from top of square on the right

Any sensible pattern with one answer option that will complete the grid correctly.

Connections with codes 3 (page 50)

Have a go

1 A black rectangle on the left side with a short horizontal line pointing to the right at the top.

2 **circle position** – A is white circle in the middle, B is white circle at the top and C is white circle at the bottom; **diamond shading** – X is black shading of the diamond shape and Y is line-shading of the diamond

 Distractors: **position** of triangle

 So pattern at the end will have a white circle in the middle and a diamond shape with diagonal-lined shading and a triangle at the top *or* bottom.

3 **position** – L is circle over line of square, M is circle inside square, N is circle outside the square; **shape** – X is small black circle at one end of curved line, Y is small straight line across one end of the curved line and Z is small white circle at one end of curved line

 Distractors: **shading** – (a) of circles (b) of squares; **angle** – of curved line; **position** – of curved line relative to the square

So pattern at the end must have a circle across the edge of a square, with a curved line passing through the circle and a small white circle at one end of the curved line.

4 **number** – (a) of zig-zag lines, where D has four, E has five and F has three, so the missing code at the top of the fourth figure is E, (b) of short lines across the zig-zag, where S has three, R has four and T has two lines

 So the missing code at the bottom of the fifth figure is T.

 Distractors: **position** – of (a) short lines across the zig-zag, (b) small circles in the figure; **shading** – of circles; **number** – of circles

Test yourself

1 b **shape** – A is for one line of symmetry in the main shape and B is for two lines; **number** – X is for three circles and Y is for four circles

 Distractors: **shading** – (a) of shape, (b) of background, (c) of circles

2 c **number** – (a) of lines, A is for five lines and B is for four lines, (b) crossing points, X is for three, Y is for four and Z is for five

 Distractors: **shading**; **line style**; **angle** – of lines within the shape

3 d **number** – (a) E is for three black dots in a row, F is for two and D is for four, (b) loops along the curly line – L is for three, M is for four and N is for five

 Distractors: **number** – of crosses; **position** – orientation of black dots; **shape** – style of endings on the coiled line

4 b **shape** – A is for circle at the top, B is for triangle at the top; **shading** – M is for black shading of overlap, N is for lined shading

 Distractors: **shape** – (a) shape of polygon in bottom part, (b) shape/style of line with arrowhead; **position** – of arrow within the figure

5 a **orientation** – G is for clockwise spiral and H for anticlockwise spiral; **shape** – S is for plain arrowhead, T for black arrowhead and R for curly arrowhead

 Distractors: **number** – of white circles; **shading** – of circle in centre of each spiral; **line style** – for each spiral

6 b **number** – (a) total number of the sides of the polygons within the square Z is for eleven, Y is for ten and X is for nine, (b) number of circles – A is for one circle, B is for three and C is for four

 Distractors: **shading** – (a) of polygons, (b) of circles; **number** – (a) of crosses, (b) of polygons

The letters in the top right-hand corners should represent one feature; the letters in the bottom right-hand corners should represent another feature.

Matrices 3 (page 52)

Have a go

1 (a) no: dashed and solid lines need reversing and bottom right half of square needs shading

 (b) correct answer

 (c) no: bottom right half of square needs to be shaded

 (d) no: bottom right half of the square needs to be shaded

 (e) no: needs double solid line along the bottom

2 (a) no: needs a line in the middle at the bottom

 (b) no: needs white circle on the left and black on the right to match the central circles in the adjacent squares, as on the bottom row

 (c) no: just two circles as in bottom row central square

 (d) no: just two circles as in bottom row central square

 (e) correct answer

3 (a) no: wrong shading – needs to be black as in diagonally opposite corner of the central square

 (b) no: needs to be a square not a circle

 (c) no: needs to be a square not a circle

 (d) correct answer

 (e) no: wrong shading of square

4 hexagon will have two solid lines inside top-left edge, a dashed arrow from top right to lower left across the middle, a black circle in the middle angle on the right and a small Z-shape in the lower right angle

Test yourself

1 d **orientation** – the V symbols are pointing the way round the grid so the next one will point down; **number** – the central row has two circles in each square; **shading** – of circles alternates along the rows

2 e **line style** – of the short diagonal line across each square follows a pattern along the rows, with each style appearing twice, so will be a double-headed arrow; **shape** – there are circles in each of the alternate squares so will have circles; **number** – two circles in alternate squares; **shading** – one black and one white circle each time

3 c **line style** – L-shapes around the centre point have same style on vertically opposite lines and one dashed one solid opposite each other on the horizontal lines; **shape** – inside the L-shape is a small shape same as shape diagonally opposite so a small circle; **shading** – of the small shape is opposite shading style to the diagonally opposite shape so a white circle; **number** – dashed line of four dashes coming from top to left and then three dashes and one dot from top into the top-right triangle is

following a decreasing number of dashes with increasing number of dots, so the line will have two dashes and two dots; **reflection** – the shapes at the left- and right-hand end of the figure are reflected in the horizontal line so there will be a Z-shape in the top-right corner

4　c　**shape** – (a) a small line shape in top left of each square within the large triangle, (b) a small shape in the bottom right of the small square within the triangle – the shape in the top left is the same as the shape in the bottom right of the next square working clockwise around the figure; **shading** – (a) lined shading for each alternate triangle, (b) shading of second triangle in each section is same as that in the diagonally opposite large triangle so missing part will have a small white triangle

5　e　**shape** – main shape in each hexagon reflected in hexagon directly opposite; **shading** – any black shaded area is striped shading in opposite figure; **line style** – the three-segmented line going round the whole hexagon alternates between solid line and dashed line

Distractor: similar shape used in adjacent figure but with alternate shading

6　c　**shapes** – the shape in the top half of one hexagon becomes the shape in the bottom half of the next but one hexagon; **line style** – horizontal lines across each hexagon are same in directly opposite hexagons; **shading** – (a) lined shading in lower half becomes black shading when on the top half in next but one hexagon, (b) black shading in shape in lower half of a hexagon becomes lined shading when shape is in the top half of next but one hexagon

Distractors: **orientation** – of the line across each hexagon remains horizontal while other elements rotate round; **shape** – patterns in central hexagon

Try it out

In your grid check that the pattern follows a clear rule that can be worked out from the information given in the grid.

Maths workout 2 (page 54)

Puzzles

1　(a)　no: the fraction shaded is $\frac{2}{9}$

(b)　no: one of four parts is shaded but the four parts are not equal, so not $\frac{1}{4}$ shaded

(c)　yes: $\frac{1}{4}$ of **c** shaded – one of four equal parts

(d)　yes: $\frac{1}{4}$ of **d** shaded – two of eight equal parts, $\frac{2}{8} = \frac{1}{4}$

(e)　no: four equal parts shaded but the whole is divided into 12 parts, so $\frac{4}{12} = \frac{1}{3}$, not $\frac{1}{4}$

2　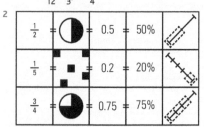

3　For example:

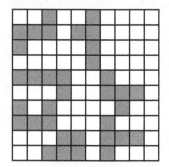

1　$5 \times 2 + 4 = 14, 14 \times 2 + 4 = 32, 32 \times 2 + 4 = 68, 68 \times 2 + 4 = 140$

2　$720 \div 2 + 2 = 362, 362 \div 2 + 2 = 183, 183 \div 2 + 2 = 93.5, 93.5 \div 2 + 2 = 48.75$

3　$1 \times 3 = 3, 3 \times 3 = 9, 9 \times 3 = 27, 27 \times 3 = 81$

4　Third term – a circle containing five short lines, one small circle and one triangle over the edge of the circle at some point; fourth term – a circle containing nine short lines, one small circle and one triangle over the edge of the circle at some point

5　Third term – a zig-zag line with nine sections, any lengths or angles; fourth term – a zig-zag line with 17 sections, any lengths or angles

6　Third term – a right-angled zig-zag with four sections each at right angles, a black spot on the first end and a white circle along the zig-zag; fourth term – a zig-zag with five sections each at right angles, a white spot on the first end and a black circle along the zig-zag

7　Fourth term – a hexagon shaded black; fifth term – seven-sided polygon shaded with oblique lines

8　

Crosswords and logic

1

1	9	3	1			3
1	3	7				2
			9	0	1	
2	2	2				
4			5	7	9	
0		2	5	5	0	

2　(1)　subtract
(2)　parallel
(3)　divide
(4)　square
(5)　percent
(6)　angle
(7)　total
(8)　perpendicular
　　　TRIANGLE

3　27: odd numbers between 25 and 33 are 27, 29, 31; of those only 27 is a multiple of 3 and 9

4　64: square numbers less than 100 are 1, 4, 9, 16, 25, 36, 49, 64, 81; take out the odd ones leaves 4, 16, 36, 64; 4, 16 and 36 are square numbers but not cubed numbers; 64 is the square of 8, $8 \times 8 = 64$, it is the cube of 4, $4 \times 4 \times 4 = 64$

5　20: prime numbers have only two factors, 1 and themselves, so the first three prime numbers are 2, 3 and 5 (1 is not a prime number) so $2 + 3 + 5 = 10$, and $2 \times 10 = 20$

Test yourself

The first cube has been rotated to a new position, shown after the arrow. One of the answer options shows the third cube, rotated in the same way as the first cube. Circle the letter beneath the correct answer. For example:

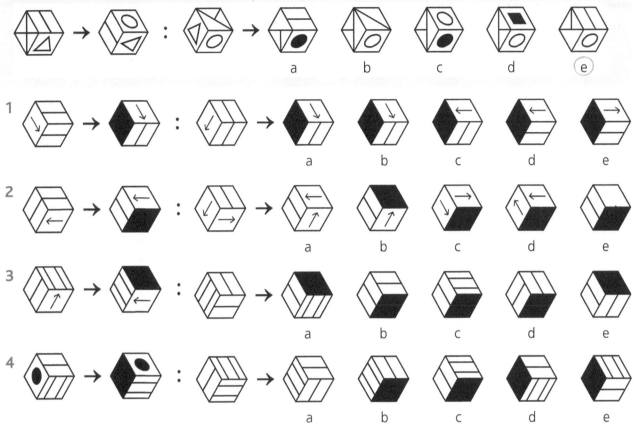

Score ☐ / 4

Try it out

Draw your own shapes and patterns on these blank cubes to make a similar question for your friend or parent to complete.

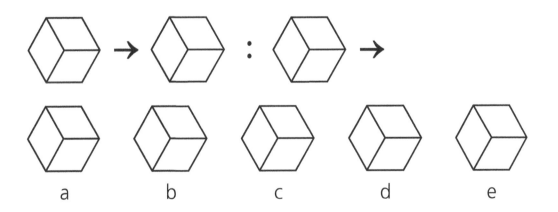

a   b   c   d   e

Translating and combining images 2

Without rotating the patterns on the left, join the dots on the right to produce the same pattern. In the first two examples, the dots have been joined to help you find the pattern.

1 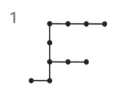  2

Now find the patterns that both appear twice in the grids on the right. Join the dots to show where the patterns are.

3    4

In the next two questions, the pattern is rotated 90 degrees anticlockwise. Find it on the grid and join the dots.

5   6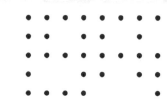

Test yourself •

The pattern of dots in the left-hand box is hidden in the dots in the right-hand box. Carefully join the dots of the pattern once you find it. For example:

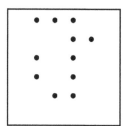 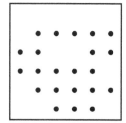 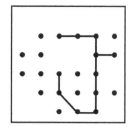

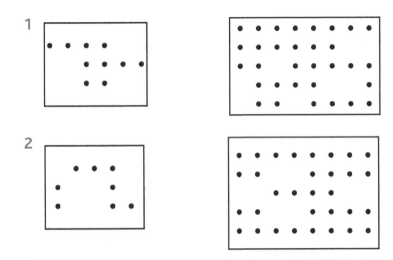

1

2

Notice that the pattern has been rotated in the next two questions.

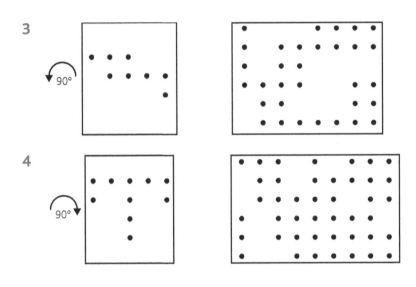

3

4

Score ☐ / 4

Try it out

Use the grid on the right to hide these patterns by adding large dots. Remember there must be only one place where they can be drawn.

1

2

Translating and combining images 3

Which of the two 3D figures on the right could be made from combining only the blocks on the left? You must not use any blocks more than once and all the blocks must be used. Some blocks may be rotated. Circle **a** or **b**.

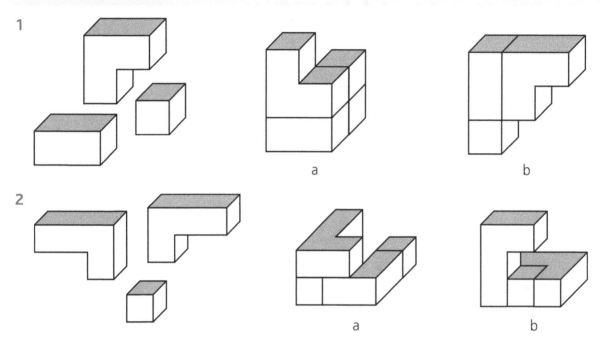

Circle all the letters of the blocks on the right that are needed to build the 3D figure on the left.

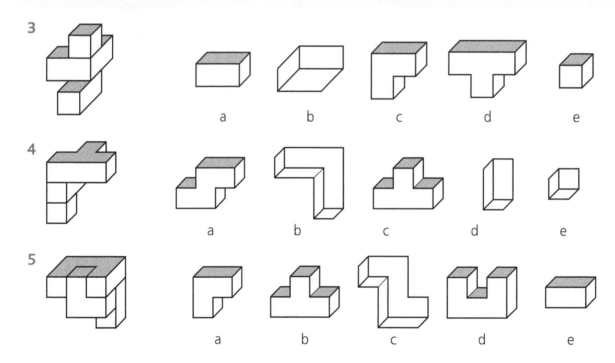

Test yourself

One group of blocks on the right has been joined together to produce the figure on the left. Some of the blocks may have been rotated. Circle the letter beneath the blocks that make up the figure on the left. For example:

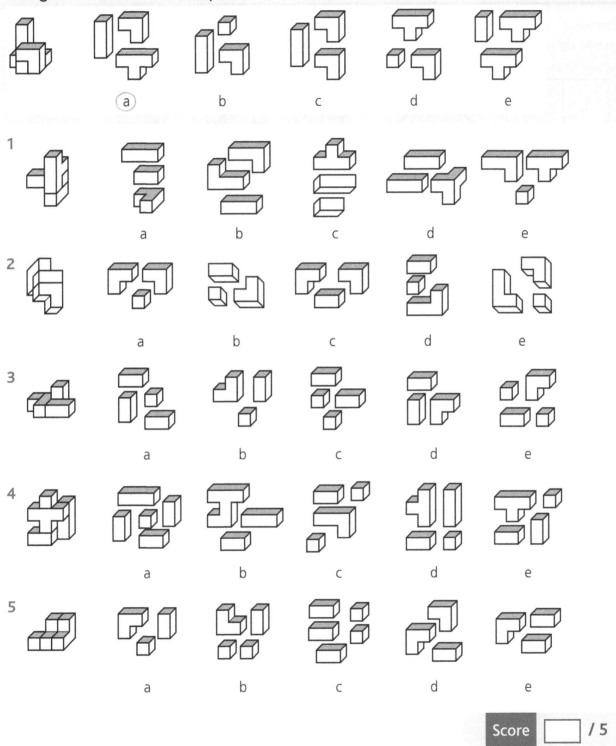

a   b   c   d   e

1   a   b   c   d   e

2   a   b   c   d   e

3   a   b   c   d   e

4   a   b   c   d   e

5   a   b   c   d   e

Score [   ] / 5

Try it out

How many different 3D patterns can you build using six 1cm cubes if each cube must have at least one face touching another cube? Try sketching the different options.

Translating and combining images 4

Each of the small squares on the right of the question fits perfectly into the diagram. Show where each of the small squares fits by writing its letter on the numbered diagram. The small squares do not rotate.

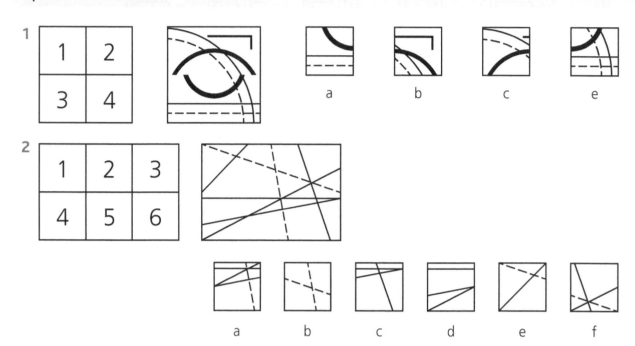

Write the letter of each small square in the blank grid to show its position in the large diagram. Draw the missing patterns in the two blank squares, k and l. The squares do not rotate.

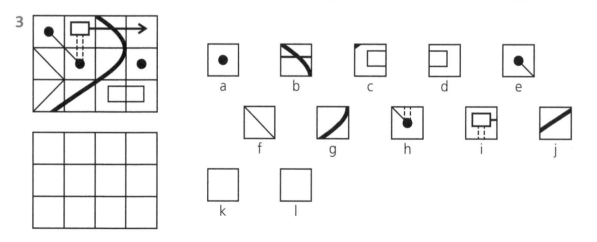

The two large diagrams have been cut into small squares. The squares have been labelled and the pieces muddled up. Decide where each piece should go in the two empty grids, using the letter name to identify the piece. The squares do not rotate. The first one has been completed for you.

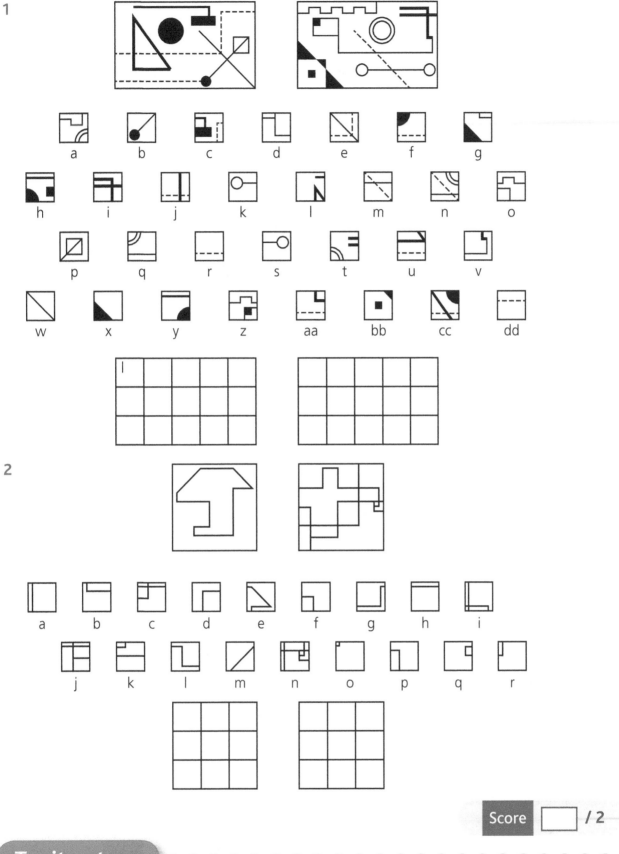

1

2

Score [ ] / 2

Draw your own pattern in a rectangle and then divide it into squares with a ruler. Copy the pattern in each square section into a separate square beside the grid. For an extra challenge you could try doing this with pictures from a magazine! Ask a friend or parent to work out the puzzle.

The Non-Verbal Reasoning papers include mathematical questions because there are many strands of mathematics that have a direct link with reasoning. The need to identify shapes and patterns, symmetry and rotation are all elements of both maths and Non-Verbal Reasoning. These quick questions will help to sharpen those mathematical skills which will, in turn, help you with Non-Verbal Reasoning questions.

Building 2D and 3D shapes

Complete these patterns where the dashed lines are lines of symmetry.

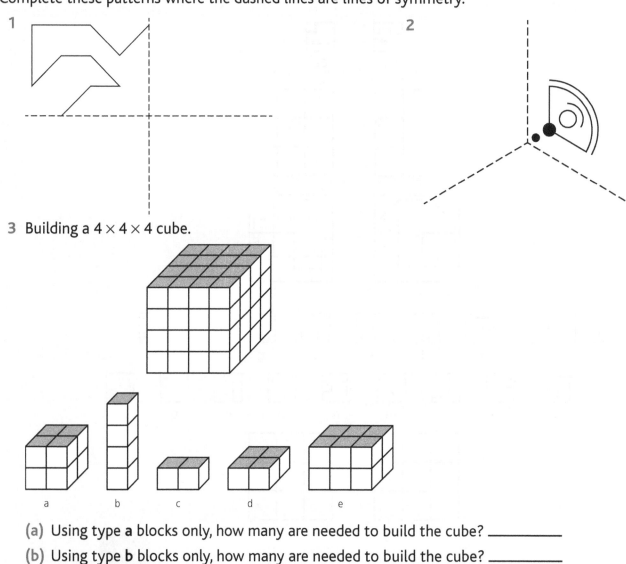

3 Building a 4 × 4 × 4 cube.

(a) Using type **a** blocks only, how many are needed to build the cube? _____

(b) Using type **b** blocks only, how many are needed to build the cube? _____

(c) Using type **c** blocks only, how many are needed to build the cube? _____

(d) Using type **d** blocks only, how many are needed to build the cube? _____

(e) If the maximum number of type **e** blocks are used, how many single cubes would be needed to complete the 4 × 4 × 4 cube? _____

Score [ ] / 7

Translating and rotating 2D shapes

1 Plot the following points on the axes below and join them up to form a polygon.

A (2,4)   B (4,4)   C (5,2)   D (1,2)

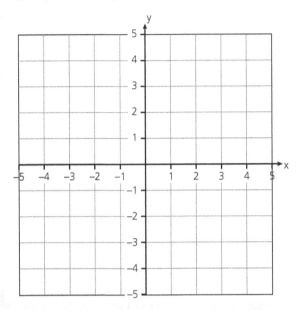

2 Reflect the polygon in the *y* axis and label the new points A', B', C' and D'.

(a) Give the new co-ordinates for   A'(,)   B'(,)   C'(,)   D'(,)

(b) Translate the original polygon by moving down five squares, and three squares to the left.

(c) Label the new points A'', B'', C'' and D''.

(d) Give the new co-ordinates for   A''(,)   B''(,)   C''(,)   D''(,)

3 Each picture on the left of the grid below has seven blank squares to its right. Redraw the picture in every blank square on the row by following the instructions along the top of the grid in sequence.

	90°	180°	45°	135°	45°	90°	135°
(a)							
(b)							
(c)							

Score ☐ / 8

37

C Codes, sequences and matrices
Connections with codes 1

Have a go

1 The first letter in these codes represents shape and the second letter represents number. Work out and write in the missing code letters.

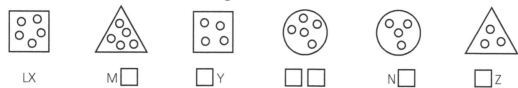

LX   M☐   ☐Y   ☐☐   N☐   ☐Z

2 The first letter in these codes represents shading and the second letter represents position. Work out and write in the missing code letters.

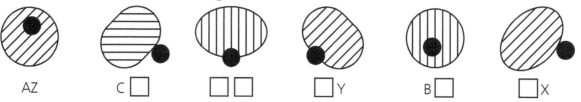

AZ   C☐   ☐☐   ☐Y   B☐   ☐X

3 Work out what each code letter represents and write the code for the last pattern.

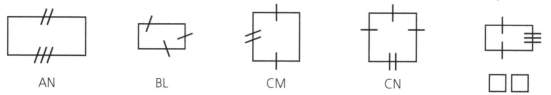

AN   BL   CM   CN   ☐☐

4 Here, there are three elements represented by codes. Work them out and then give the code for the last pattern.

AMY   BLZ   BNX   CMZ   ☐☐☐

5 Work out the missing codes and then write down two features, or elements, that are distractors that do not form part of the necessary rules for working out the coding.

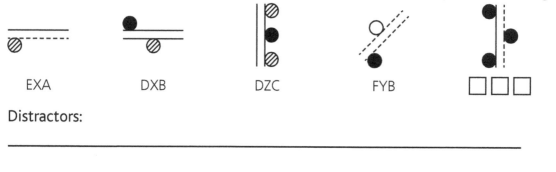

EXA   DXB   DZC   FYB   ☐☐☐

Distractors:

_____

_____

Test yourself

Each letter represents an individual feature in the picture next to it. Work out which feature is represented by each letter. Apply the code to the picture in the box and circle the letter beneath the correct answer code. For example:

 SUW

 TVX

 TUY

SVZ

TVZ	SUY	SVX	SUW	TUZ
a	b	c	d	(e)

1
AY

BX

BZ

CY	AZ	BY	AX	CZ
a	b	c	d	e

2
AXM

BYM

CXN

AXN	BZN	BXM	AZN	CZM
a	b	c	d	e

3
AXL

 BYM

AZM

BZL	BYL	AZL	AXM	BZM
a	b	c	d	e

4
DLY

ELZ

DNX

FNY

DLZ	FMZ	EMY	FMX	EMZ
a	b	c	d	e

Score ___ / 4

Try it out

Draw a set of patterns in the first three squares and apply a code letter to each feature of your pattern. Then draw a fourth pattern including a mixture of features to create a unique code. See if a friend or parent can work out the code of the fourth pattern.

Sequences 1

Draw the next two pictures in these sequences.

1

2

Complete the next two sequences by drawing in the missing pictures.

3

4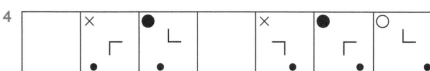

5 Identify the different elements within this sequence.

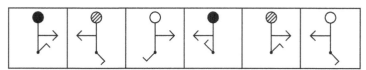

    (a) What individual element is alternating? _____

    (b) What element is rotating 90 degrees clockwise? _____

    (c) There is one element which is made up of a repeating
        pattern of three. Draw the set of three.

The five boxes on the left show a pattern that is arranged in a sequence. Choose the answer option that completes the sequence when inserted in the blank box. Circle the letter beneath the correct answer. For example:

                a   b   c   d   (e)

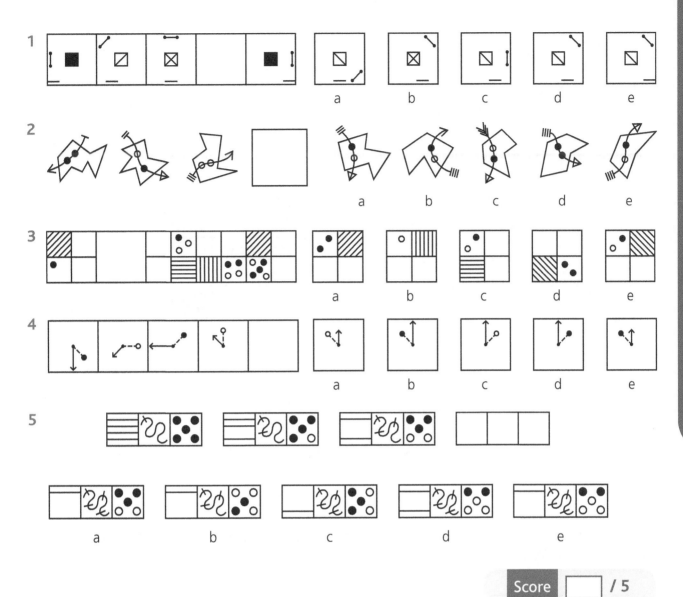

1

2

3

4

5

Score ☐ / 5

Try it out

Make up a sequence of four pictures starting with the pattern below and see if a friend or parent can draw the fifth diagram.

a   b   c   d   e

Matrices 1

Draw the missing pattern in the square provided and explain your answer.

1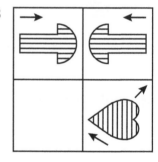

Reason: _____

_____

_____

_____

2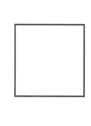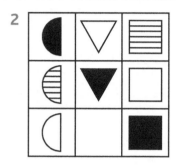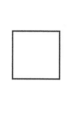

Reason: _____

_____

_____

_____

Draw the missing pattern in the square provided and draw lines over the grid to show any lines of symmetry.

3

4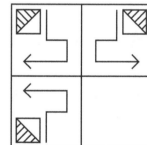

Complete the next two grid patterns where the dashed lines drawn are lines of symmetry.

5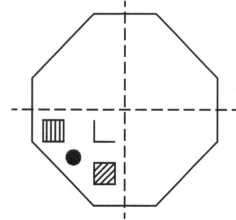

6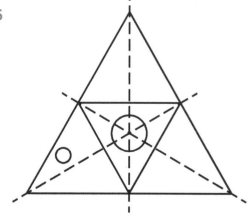

Test yourself

One of the options on the right completes the pattern in the grid on the left. Circle the letter beneath the correct answer. For example:

   ⓐ   b   c   d   e

1

    a   b   c   d   e

2

    a   b   c   d   e

3

    a   b   c   d   e

4

    a   b   c   d   e

Score ☐ / 4

Try it out

Draw a pattern in the grid below where shape, number and shading change in some way. Remember, you can use symmetry, columns or rows to create your pattern. Leave a blank square and give five answer options. Challenge a friend or parent.

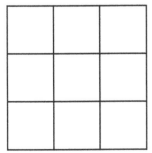

                        a   b   c   d   e

Connections with codes 2

From the pictures shown, work out what the codes mean. Then draw another, different but related, picture and write its code.

1   2

    AY   BZ   BX   DAX   EAY   DBZ

Complete these pictures by adding one or more elements so that the codes written beneath them could be applied.

3 4

    AY   BX   CY   BZ   LRY   LSZ   MTZ

Each letter represents an individual feature in the picture next to it. Work out which feature is represented by each letter. Apply the code to the picture in the box and circle the letter beneath the correct answer code. For example:

 SUW

 TVX

 TUY

 SVZ

     TVZ   SUY   SVX   SUW   TUZ

                     a   b   c   d   (e)

1

    AY   BX   BZ   AX

         AZ   BY   BX   AY   AX

         a   b   c   d   e

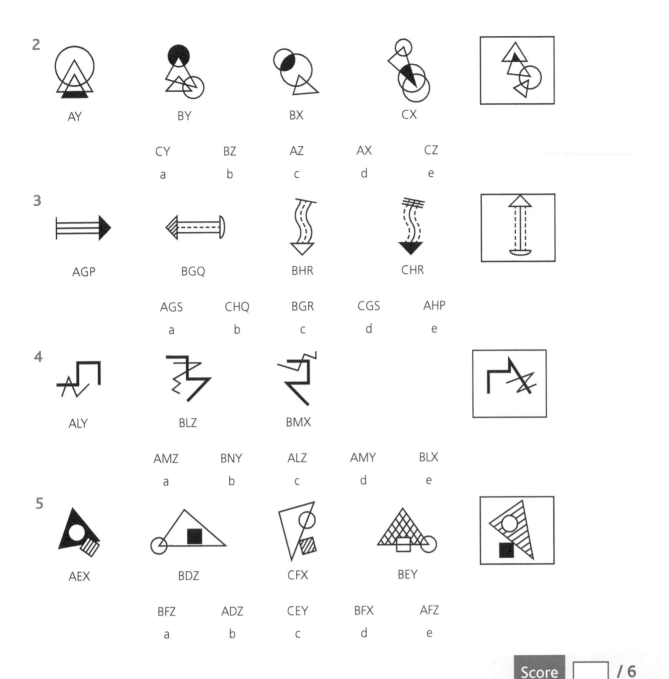

2

AY BY BX CX

CY BZ AZ AX CZ
a b c d e

3

AGP BGQ BHR CHR

AGS CHQ BGR CGS AHP
a b c d e

4

ALY BLZ BMX

AMZ BNY ALZ AMY BLX
a b c d e

5

AEX BDZ CFX BEY

BFZ ADZ CEY BFX AFZ
a b c d e

Score [ ] / 6

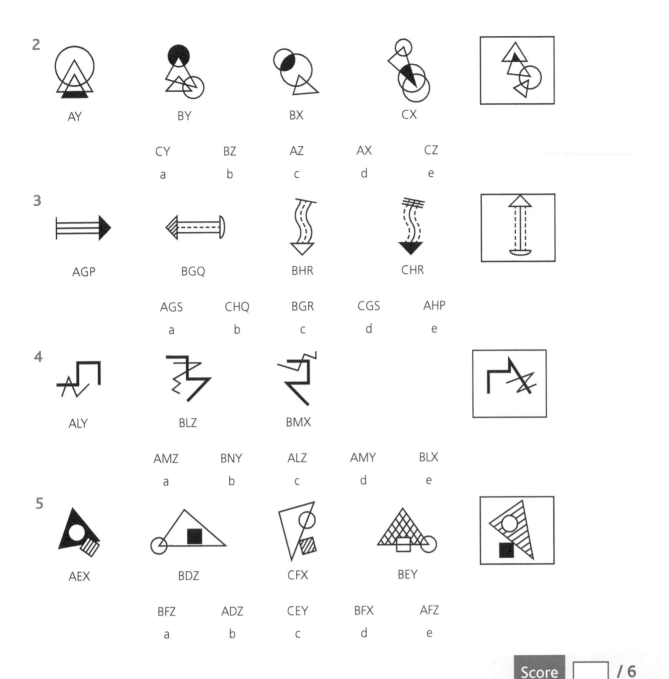

Try it out

Draw your own five patterns, matching them to three code letters and writing out the codes for the first four patterns.

The **first** code letter should relate to a **changing shape**, the **second** code letter to **different quantities of a single shape** and the **third** code letter to **variations in shading of one element** of your pattern. Ask a friend or parent to work out the code for the final pattern.

Sequences 2

1 A sequence has been started. In the top box, draw the next two pictures that might follow if it was a repeating sequence. In the bottom box, draw the next two pictures that might appear if the sequence was made up of two alternating sequences.

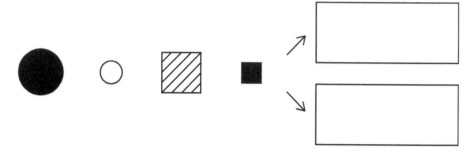

2 In this sequence, draw the next two pictures that might appear if it was a reflective sequence (a sequence of five pictures that make a symmetrical pattern) in the top box. Then, in the bottom box, draw the next two pictures if the sequence was a continuing sequence.

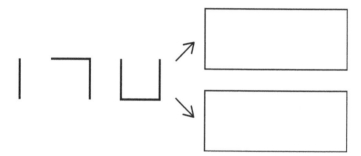

3 List three elements that are changing between each picture in the next two sequences.

_____

_____

_____

_____

The five boxes on the left show a pattern that is arranged in a sequence. Choose the answer option that completes the sequence when inserted in the blank box. Circle the letter beneath the correct answer. For example:

            a   b   c   d   (e)

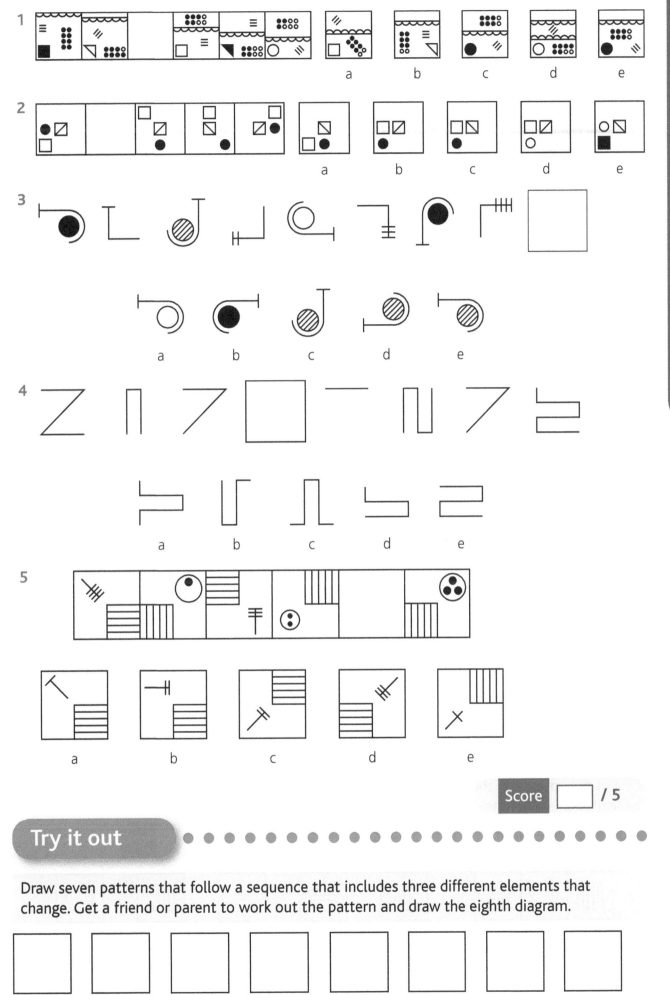

1

    a   b   c   d   e

2

    a   b   c   d   e

3

    a   b   c   d   e

4

    a   b   c   d   e

5

    a   b   c   d   e

Score [ ] / 5

Try it out

Draw seven patterns that follow a sequence that includes three different elements that change. Get a friend or parent to work out the pattern and draw the eighth diagram.

Matrices 2

1 Draw the missing pattern in the blank square of this grid.

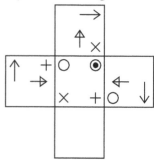

2 In this grid there are three missing elements spread across the sections – identify the missing elements and draw them in to complete the pattern.

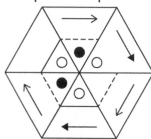

3 Draw the missing patterns from this extended grid (they are numbered 1 to 4) in the blank squares on the right.

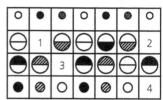

4 Identify the missing pattern section, circle the answer option letter, and write the reason for your choice.

Because ... _____

_____

One of the options on the right completes the pattern in the grid on the left. Circle the letter beneath the correct answer. For example:

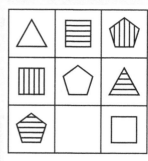

a   b   c   d   e

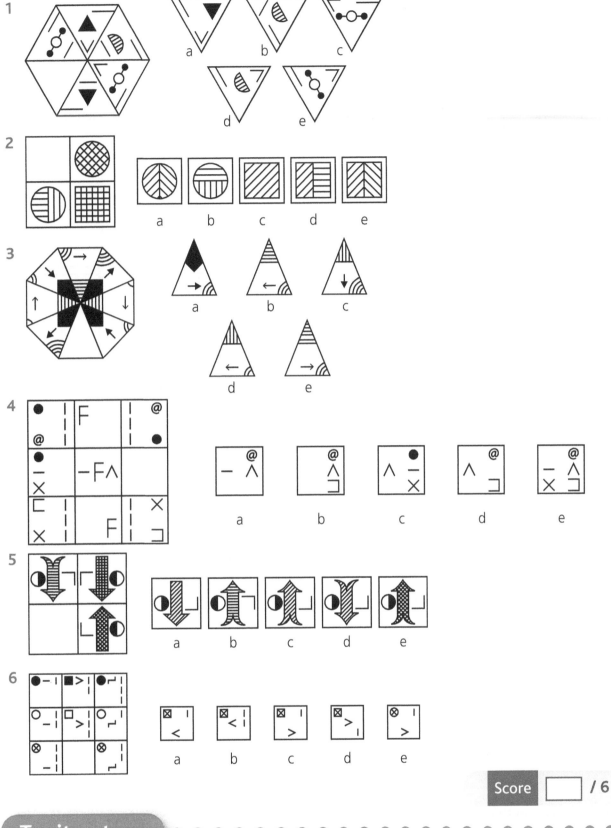

Score ☐ / 6

Try it out

Draw a pattern in the grid below. Leave the last triangle blank. Give five possible options for the last triangle. Ask a friend or parent to identify the missing pattern.

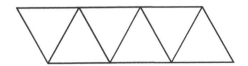

Connections with codes 3

Draw the pattern that matches the codes given in the last picture.

1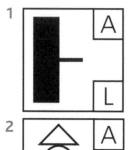

2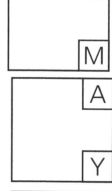

3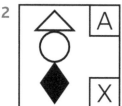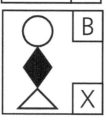

4 Identify and fill in the missing code letters. List the elements that are distractors.

  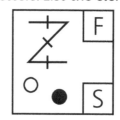  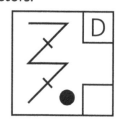

Distractors: _____

_____

_____

The two letters in the small boxes each represent a feature of the shapes in the large box. Work out which feature is represented by each letter and apply the code to the final box. Circle the letter beneath the correct answer code. For example:

                                                         a   b   c   (d)   e

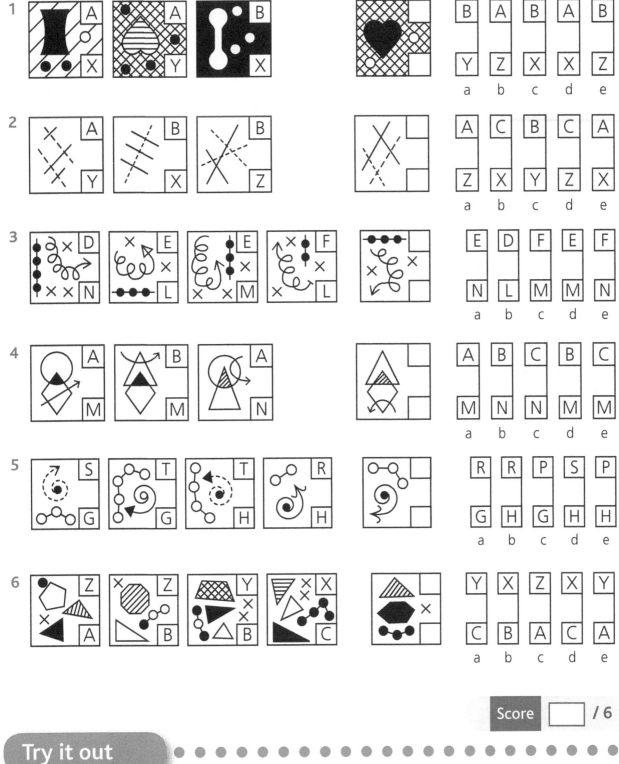

Score ☐ / 6

Try it out

Draw diagrams in these squares that could go with the coding given.

Matrices 3

Circle the letter next to the answer option that completes the grid and give one reason why each of the other options is not correct.

1

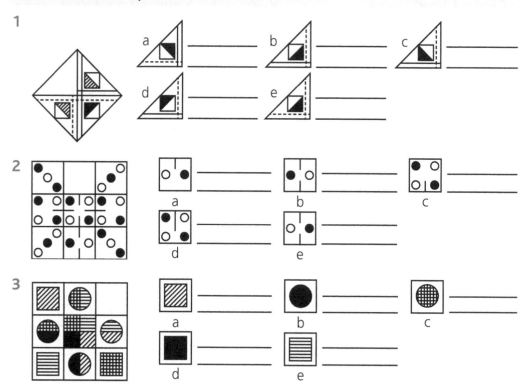

2

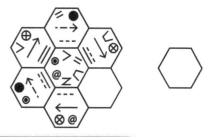

3

4 Now draw the missing pattern in the blank hexagon.

One of the options on the right completes the pattern in the grid on the left. Circle the letter beneath the correct answer. For example:

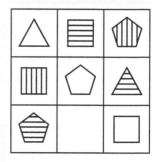

(a)   b   c   d   e

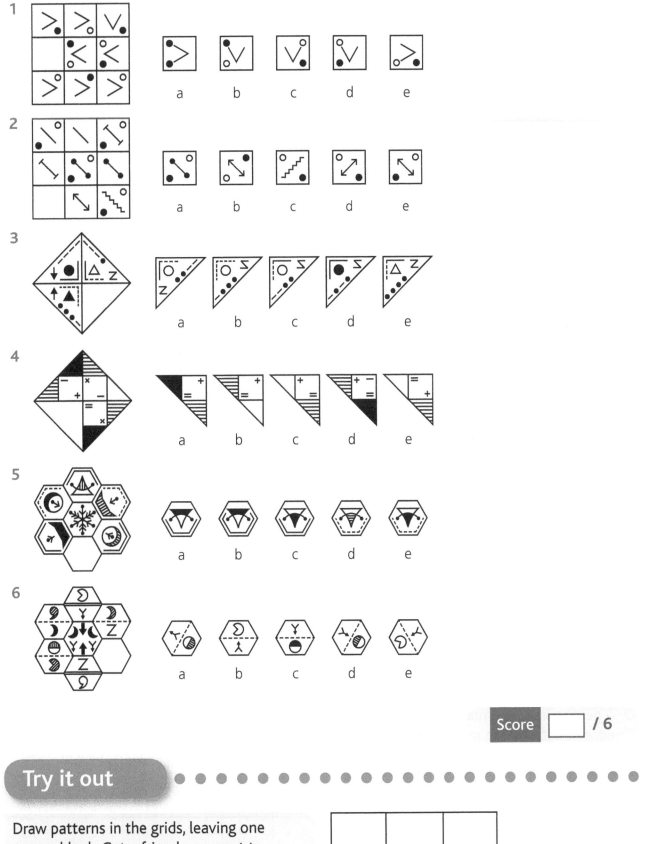

1

2

3

4

5

6

Try it out

Draw patterns in the grids, leaving one
square blank. Get a friend or parent to
work out the missing pattern. Try to use
two different types of pattern, for example
repeating and increasing patterns.

Maths workout 2

Many Non-Verbal Reasoning problems make use of mathematical skills and knowledge, so these pages contain some questions and puzzles to consolidate your mathematical skills, vocabulary and ideas. Keeping your maths skills sharp will help you to solve Non-Verbal Reasoning questions more quickly!

Puzzles

1   Which of the following shapes have a quarter shaded? Circle their letters.

a   b   c   d   e

2   In this table, fractions are presented numerically and diagrammatically. Complete the table to make it correct.

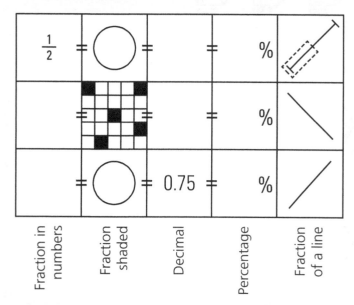

3   A pentomino tile is made up of five squares joined together to form a shape. At least one edge of each square must touch the edge of another square.

Here are two examples of pentomino tiles:

Can you draw seven different pentomino tiles in the grid below?

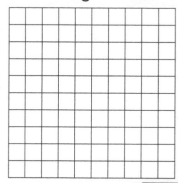

Draw in any lines of symmetry on the pentomino tiles that you draw.

Score ☐ / 3

Sequences

Follow these rules to give the next four numbers in these sequences. Remember to apply the rule each time.

1 $\times 2$ $+4$ 5   _____ _____ _____ _____

2 $\div 2$ $+2$ 720   _____ _____ _____ _____

3 $\times 3$ 1   _____ _____ _____ _____

Use the rules to work out and then draw the next two diagrams in these sequences, where n is the next term number and x = the number of lines in the diagram. Remember to notice what other features are present and to include them each time.

4 $n = 2x - 1$

5 $n = 2x - 1$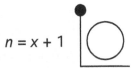

6 $n = x + 1$

7 $n = x + 1$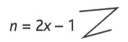

8 Draw the first three diagrams in a sequence where the rule can be described as: a zig-zag line, with an additional line in the zig-zag each time; a circle, alternating black and white, at one end and an increasing number of short lines across the other end, starting with two short lines; and with the orientation of the zig-zag alternating between vertical and horizontal.

Score _____ / 8

Crosswords and logic

Work out the answers to the questions and fill in the crosswords.

1

Across	Down
1 2003 – 72	**3** 789 – 468
4 $11^2 + 4^2$	197 + 53
5 $10^3 - 99$	
6 74×3	
7 123 + 456	
8 510 – 5	

2

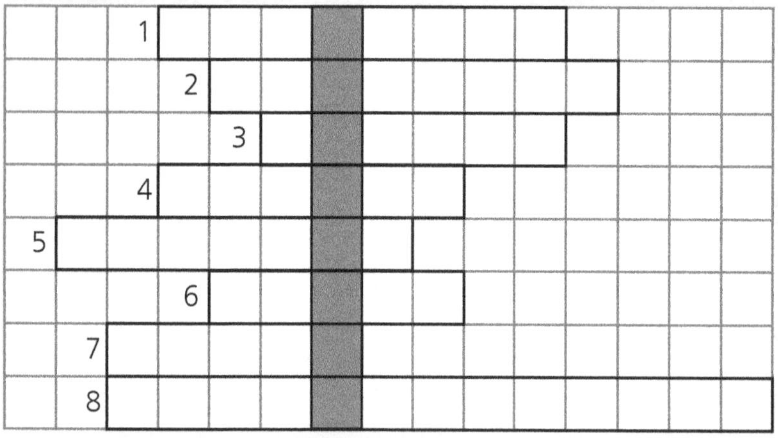

Fill in the missing words:

1 Another word for take away.

2 A word used to describe two straight lines which are an equal distance apart from each other for their whole length; they never meet.

3 The mathematical term meaning sharing out.

4 The name of a four-sided shape where the four sides are all the same length and the four angles are all 90 degrees.

5 The name for a fraction which is out of 100.

6 What can be described by the words 'acute' or 'obtuse'?

7 Another word for adding all together.

8 The name given to a line which is drawn at right angles to another line.

What word is made by the letters in the vertically shaded column of this grid?

_____

I am an odd number, greater than 5^2 and less than 3×11. I am a multiple of 3 and 9. What

number am I? _____

I am a square number and a cubic number. I am an even number less than 100. What

number am I? _____

I am twice the total of the first three prime numbers. What number am I? _____

Score ☐ / 5